Very Large Floating Structures

Spon Research

publishes a stream of advanced books for built environment researchers and professionals from one of the world's leading publishers.

Published

Free-Standing Tension Structures:
From Tensegrity Systems to
Cable-Strut Systems
978–0–415–33595–9
B.B. Wang

Performance-Based Optimization of
Structures: Theory and applications
978–0–415–33594–2
Q.Q. Liang

Microstructure of Smectite
Clays & Engineering Performance
978–0–415–36863–6
R. Pusch and R. Yong

Procurement in the Construction
Industry
978–0–415–39560–1
W. Hughes et al.

Communication in Construction
Teams
978–0–415–36619–9
C. Gorse and S. Emmitt

Concurrent Engineering in
Construction
978–0–415–39488–8
C. Anumba

People and Culture in Construction
978–0–415–34870–6
**A. Dainty, S. Green and
B. Bagilhole**

Very Large Floating Structures
978–0–415–41953–6
**C.M. Wang, E. Watanabe and
T. Utsunomiya**

Forthcoming

Innovation in Small Construction Firms
978–0–415–39390–4
P. Barrett, M. Sexton and A. Lee

Construction Supply Chain Economics
978–0–415–40971–1
K. London

Location-Based Management System for
Construction: Improving productivity
using flowline
978–0–415–37050–9
R. Kenley and O. Seppanen

Employee Resourcing in Construction
978–0–415–37163–6
A. Raiden, A. Dainty and R. Neale

Tropical Urban Heat Islands: Climate,
Buildings and Greenery
978–0–415–41104–2
N.H. Wong and C. Yu

Very Large Floating Structures

Edited by
C.M. Wang, E. Watanabe,
and T. Utsunomiya

Routledge
Taylor & Francis Group

LONDON AND NEW YORK

First published 2008 by Taylor & Francis

2 Park Square, Milton Park, Abingdon, Oxfordshire OX14 4RN
52 Vanderbilt Avenue, New York, NY 10017

Routledge is an imprint of the Taylor & Francis Group, an informa business

First issued in paperback 2019

British Library Cataloguing in Publication Data
A catalogue record for this book is available from the British Library

Library of Congress Cataloging in Publication Data
Wang, Chien-ming.
    Very large floating structures / C.M. Wang, E. Watanabe and
T. Utsunomiya.
        p. cm.
    Includes bibliographical references and index.
        1. Offshore structures. I. Watanabe, E. II. Utsunomiya, T. III. Title.

TC1665.W36 2007
627'.98–dc22   2007028603

ISBN13: 978-0-415-41953-6 (hbk)
ISBN13: 978-0-367-38840-9 (pbk)

Contents

Notes on contributors

 **Chien Ming Wang** is Professor of Civil Engineering at the National University of Singapore and the Deputy Head for the Engineering Science Programme. Prof. Wang is a Chartered Structural Engineer, a Fellow of the Institution of Engineers Singapore and a Fellow of the Institution of Structural Engineers (UK). He is presently the Chairman of the IStructE Singapore Division. His research interests are in the areas of structural stability, vibration, optimization, plated structures, and Mega-Floats. He is the author or co-author of over 300 technical papers, 3 books – *Vibration of Mindlin Plates, Shear Deformable Beams and Plates: Relationships with Classical Solutions*, and *Exact Solutions for Buckling of Structural Members* and co-editor of two volumes on *Analysis and Design of Plated Structures*. Moreover, he is the Editor-in-Chief of the *International Journal of Structural Stability and Dynamics*, the *IES Journal A: Civil and Structural Engineering* and an Editorial Board Member of *Engineering Structures*.

 **Brydon T. Wang** is a PhD Candidate at the University of Melbourne, having received his Bachelor of Architecture from the same university in 2005. He has worked in the United Kingdom, Singapore, and in Australia with Arkitek Tenggara, Minifie Nixon Architects, and Monarchi. His research interests lie in the confluence of Architecture and Public Policy, and include Utopian Proposals, Floating Settlements, Very Large Floating Structures (VLFS), Informal Settlements, and Global Urban Projects.

Seiya Yamashita graduated from Yokohama National University in 1968. He joined the Research Institute of Ishikawajima-Harima Heavy Industries Co. Ltd. and engaged in the study on propulsive and seakeeping performances of ships. Later he worked on the wave-induced forces and motions for offshore structures such as box-shaped floating vessels, semi-submersible drilling rigs, and floating breakwaters. He received his Doctor of Engineering in 1986 from the University of Tokyo for his study on hull configuration for offshore structures on which wave-exciting forces are not exerted in waves. From 1995 to 2000 he joined the Mega-Float project in Japan as a member of the research on hydro-elastic analysis for a VLFS. He is now a part-time manager of ship and marine technology department of IHI Co. Ltd.

Tomoaki Utsunomiya is Associate Professor of Civil and Earth Resources Engineering at Kyoto University, Japan and has undertaken education and research in the areas of Offshore Structures including Very Large Floating Structures (VLFS) and Floating Bridges. He has published about 170 papers and books in these areas. His main research interests include hydroelastic analysis of Very Large Floating Structure. Part of his major contributions in this area are "An eigenfunction expansion-matching method for analyzing the wave-induced responses of an elastic floating plate," *Applied Ocean Research*, 17 (1995); "Wave response analysis of a box-like VLFS close to a breakwater," Proceedings of the 17th OMAE (1998); and "Fast multipole method for wave diffraction/radiation problems and its applications to VLFS", *Int J Offshore Polar Eng*, 16 (2006).

Masahiko Fujikubo is Professor at the Division of Structural Engineering, Department of Social and Environmental Engineering, Hiroshima University. He graduated from Osaka University in 1979, and received his MSc in 1981 and PhD in 1988 from Osaka University. His research areas are the ultimate strength and structural reliability of ships and offshore structures, and non-linear structural analysis. He stayed in the Norwegian University of Science and Technology from 1988 to 1989 to do research on the assessment of ductile fracture of offshore tubular structures. He was engaged in the Mega-Float project in Japan from its initial phase as a technical advisor for structural design and analysis. He published a textbook *Structural Design of Very Large Floating Structure* from Seizando (in Japanese) in 2004. He is now a Director of Japan Society of Naval Architects and Ocean Engineers (JASNAOE), a Member of ISSC2009 Technical Committee "Ultimate Strength," and an Editorial Board Member of the *Journal of Marine Science and Technology*.

Shigeru Ueda graduated from Kyoto University in 1967, received his MSc in 1969 and Doctor of Engineering in 1985 from Kyoto University. He then joined the Ministry of Transport (MOT) and engaged in the design of container berth in Port of Kobe. He moved to the Port and Harbour Research Institute MOT, and did research work on design of offshore structures. Major works include berthing and mooring of ships, motions and moorings of floating structures, design of pile type offshore structures, earthquake-resistant design and reliability design of port and harbour structures. He stayed in HRS Wallingford from 1975 to 1976 and studied ship motions and analysis. He moved to Tottori University in 1994 and continues with the aforementioned research studies as well as teaches Structural Mechanics, Probability and Statistics, Dynamic Response Analysis and Offshore Structural Engineering. He is a Fellow Member of Japan Society of Civil Engineers, Professional Engineer (Civil Engineering), and Executive Professional Civil Engineer (JSCE Infra Structure Design).

Tetsuya Hiraishi is the Head of the Wave Division, Maritime Environment and Hydraulic Engineering Department, Port and Airport Research Institute. He graduated from Kyoto University in 1980, and received his MSc in 1982 from Kyoto University. Then he joined the Port and Harbour Research Institute, Ministry of Transport (present Port and Airport Research Institute). He has been mainly engaged in developing a design code for port facilities in directional random sea states. For his contribution to the design of a directional random wave maker and its applications, he obtained Doctor of Engineering in 1992 from Kyoto University. His major research interests are the implementation of mitigation tools for tsunami, storm surge and high waves, the estimation and reduction of wave overtopping at seawalls and the countermeasure against harbor agitation due to long period waves. He belongs to the Japanese Society of Civil Engineers (JSCE) and the International Society of Offshore and Polar Engineers (ISOPE). In 2005, he was given the ISOPE Award for his activity as a member of the Board of Directors, 2003–2005.

Shigeo Ohmatsu graduated from Kyushu University in 1970. After graduation, he joined the Ship Research Institute of Ministry of Transport (SRI, MOT) and engaged in research on ship motion in waves. He received his Doctor of Engineering in 1980 from Kyushu University where he worked on wave radiation/diffraction problems in time domain of floating bodies. He stayed at École Nationale Supérieure Mécanique de Nantes

(France) from 1980 to 1981 where he studied wave energy absorbing theory using floating devices. In 1984, he moved to the Ocean Engineering Division in SRI and worked on prediction methods for external load acting on offshore structures. He is a coauthor of textbook *Mega-Float Offshore Structure* (in Japanese) published in 1995. He also engaged in the Mega-Float project as a Director of Ocean Engineering Division of SRI, MOT as a co-research partner of Technological Research Association of Mega-Float. He is now a senior researcher in National Maritime Research Institute (NMRI, former SRI).

Eiichi Watanabe is a Professor Emeritus, Kyoto University and the Chairperson of Board of Directors of Foundation of Osaka Regional Planning Institute. He graduated from Kyoto University in 1964, received his first MSc in 1966, from Kyoto University, second MSc in 1968, and PhD in 1969 from Iowa State University as a Fulbright grantee and Doctor of Engineering from Kyoto University. Professor Watanabe served as a Vice President of JSCE from 2004 to 2005 and is serving as a Vice President of IABMAS, member of Board of Directors of JSSC and Chairman of Bridge Asset Management for Aomori Prefecture and City of Osaka. He is a member of the European Academy of Sciences. His research interests are in the areas of steel structures, buckling, earthquake resistant design, reliability, maintenance and durability of steel bridges, corrugated steel webs, steel cables, creep and relaxation of cables, offshore structures, and floating bridges. He has written over 440 scientific publications besides approximately 30 books including *Structural Mechanics I* and *II*, Maruzen, 1999 and 2000, respectively (in Japanese), *Encyclopedia of Bridges*, Blue Backs, Kohdansha, 1991 (in Japanese). He has edited *Theoretical and Applied Mechanics* 1996, Northholland, 1997 and *Theoretical and Applied Mechanics*, Vol. 50, Science Council of Japan, 2001.

Hideyuki Suzuki is Professor of Department of Environmental and Ocean Engineering, University of Tokyo. He graduated from University of Tokyo in 1982, and received his MSc in 1984 and PhD in 1987 from University of Tokyo. He stayed in the University of California at Berkeley from 1988 to 1989 and studied structural control. His research areas are analysis and design of dynamic behavior of ocean structural system such as VLFS, floating wind turbine and riser. He published a textbook on *Structural Design of Very Large Floating Structure* from Seizando (in Japanese) in 2004. He was the Chairman of ISSC2006 Special Task Committee "Very Large Floating Structure" and an Editorial Board Member of the *Journal of Ocean Engineering*.

Preface

The ocean as a frontier for colonization has a longevity that stems from mankind's long-standing attraction to the ocean. Covering 70% of the Earth, the ocean provides a means to alleviate demands on coastal land pressure; land scarcity; avail renewable energy sources and new materials; increase food production and even to store carbon dioxide in order to mitigate global warming. As close to 50% of the industrialized world now lives within a kilometer of the coast, the demand on land resources and space is beginning to approach a critical stage as the population of the world continues to expand at an alarming rate. There is need for a sustainable and environmentally friendly development. Technological innovations that promote stewardship of the Earth's resources, especially the ocean, are vital for mankind's survival in the next millennium. One of these environmentally friendly innovations to arise in recent times is the concept of Very Large Floating Structures (VLFS) – a technology that allows the creation of artificial land from the sea without destroying marine habitats, polluting coastal waters, and altering tidal and natural current flow.

The first of its kind, this book (comprising nine chapters) provides a comprehensive treatment of the subject on VLFS. Each chapter covers an important component and area that are essential to the concept, analysis, design, construction, and maintenance of VLFS. The contributors are experts in the topic assigned to them. The book begins with some thoughts on the colonization of ocean space and an introduction to VLFS technology and their gradual appearance in the waters of developed coastal cities and countries with coastlines. Their presence is largely due to a severe shortage of land and the sky-rocketing land costs in recent times. Chapter 2 introduces readers to the wave phenomenon and wave properties which are essential for estimating the loading on the VLFS as well as to model the structure-fluid interaction. Chapter 3 deals with the hydroelastic analysis of the VLFS because of its flexibility to deform under the action of waves. This unique characteristic differentiates a VLFS from ships or semi-submersibles which undergo rigid body motions in waves. Chapter 4 presents the analysis of the VLFS from a global treatment of the structure and zooming in to the cell level of the floating structure. In Chapter 5, the station-keeping system of the VLFS is discussed. This component of the VLFS is important in keeping the huge structure from drifting dangerously in the coastal waters or harbours in which the VLFS is built. Breakwaters are needed when the VLFS are constructed in sea states that have significantly high waves. These breakwaters reduce the wave forces impacting on the floating structure. Chapter 6 addresses the analysis and design of breakwaters. Chapter 7 discusses various experimental verifications that are needed to calibrate the simulation models and provide insight into the actual flow of water through the VLFS as well as to determine the drift forces for the mooring systems. In Chapter 8, the anti-corrosion systems and maintenance of the VLFS for long-term safety and structural integrity will be described. The final chapter reports on the research and developments on the VLFS with emphasis on the Mega-Float, a 1-km long floating test runway, which was documented as the World's largest floating man-made island in the Guinness Book of Records.

The book also contains many illustrative photographs, drawings, pertinent equations describing the mathematical models for analysis, practical insights, and a large body of references on VLFS. It is hoped that this book will be a useful VLFS reference source to professional engineers, academics/researchers, architects, naval architects, and graduate students working in the offshore, marine, and structural engineering areas.

Chapter 1

Colonization of the ocean and VLFS technology

C.M. Wang and B.T. Wang

> *Ocean*: a body of water occupying about two-thirds of a world made for man who has no gills.
>
> (Ambrose Gwinett Bierce)

1.1 Colonization of ocean space

The past 50 years was marked by an increase in the movement of populations to coastal regions such that close to half of the industrialized world now lives within a kilometer of the coast (Healy and Hickey 2002). The accompanying growth of urban development in such areas has seen many countries like Japan, Dubai, Singapore, The Netherlands and Monaco expand seawards in a bid to ease demands on coastal land resources.

The scarcity of land is one line of argument for the colonization of ocean space. A second line of argument relates to lifestyle choices. The 1992 editorial of Aquapolis 4 drew attention to Australia where the majority of the population lives along the southern and eastern coasts. Although land pressure does not appear to be as great in Australia, *The Sunday Age Magazine* (April 18, 2004) noted a current trend of "sea change" that has produced thousands of suburban-style subdivisions around "erstwhile coastal villages." This new-style suburban sprawl forced Victorian planning minister, Mary Delahunty, to announce new boundaries to limit this rapid coastal development in January 2004 (Houston 2004).

With populations in coastal areas set to increase, coastal resource management becomes critical (Healy 1995). The Floating Structures Association of Japan has lobbied supplementing urban sprawl inland with expansion outward into the sea through the employment of Very Large Floating Structures (VLFS) pontoon technology that will reduce the impact on the environment. Such technology will not only provide relief from population pressures, but, as pointed out in Aquapolis 4, enhances the intimate relationship between coastal cities and their offshore environments.

The third line of argument relates to the colonization of the ocean for minerals, hydrocarbons, renewable energy sources (such as waves and wind) and increased food production. The need for arable land utilized in terrestrial food production often competes with the needs of forests and natural ecosystems, with potential impacts on global climate change (Matsuda et al. 1999). Proponents of open-ocean ranching believe that a Blue Revolution (The marine version of the Nobel Prize-winning Green Revolution that enabled many developing countries to produce more food to feed their populations.) will provide "increased protein production while improving the environment."

1.2   Historical context

The ocean as a frontier for colonization has a longevity that was rekindled during the 1960s and 1970s when the technological positivism of the previous decade met with the prevailing anxieties of overpopulation and environmental degradation. Societal anxieties fuelled by the oil crises of 1973 and 1979, and the discovery of the depletion of the ozone layer led to a strong desire to colonize hostile environments such as the deserts, the poles, the sea and even outer space. This generated a plethora of images, written work, utopian proposals, and experimentation. However, many of the proposals did not materialize and were accused of being just "paper architecture" (Thomsen 1994).

In the 1960s and 1970s, military, economic and nationalist objectives often supported the exploration of hostile environments, availing strong development in technologies and techniques employed in the fields of mining, defense, oceanography, transportation and aerospace exploration. These enabled the engineering requirements and materials necessary for the colonization of hostile environments. Of these hostile environments, the sea offers the closest climatic condition to that of land and as such, allows easy technological crossovers from the long-established shipping industry and other land-bound industries for transport, marine culture and construction infrastructure. With the establishment of the 1982 UN Convention of the Law of the Sea (UNCLOS) and the formation of organizations such as the International Society of Offshore and Polar Engineers (ISOPE) and the International Seabed Authority (ISA), the twenty-first century, as iterated by Tsutomu, is set to be "the century of ocean" (Fuse 2003).

There is a long global history of land reclamation. Projects such as the "bordos"—artificial islands built in the Epiclassic period (*c.* AD 650–900) in the Chignahuapan Marsh in Mexico (McClung de Tapia and Sugiura 2002); large-scale reclamation work by the Tokugawa Feudal Government in Nagasaki between the seventeenth and nineteenth centuries as well as land reclamation projects carried out during Roman Times in the estuaries and coasts of the United Kingdom (Davidson et al. 1991); are but a few examples.

Concurrent with the utopian proposals of the time, the 1960s saw a number of artificial islands declared independent nations. The first of these, *Sealand*, was founded in 1967 on an abandoned World War II gun platform in the North Sea, and has survived to the present day. This abandoned gun platform is currently home to the data co-location site, *HavenCo.com*. Existing offshore installations and artificial islands, together with the marine cities of the future were studied by an international panel of scientists at the symposium on *Ocean Cities '95* in Monaco. At the time, French engineer and futurologist Thierry Gaudin highlighted that technology had progressed to a level where living on or under the sea was within reach, diminishing the primacy of land reclamation as the default means of expansion into water bodies (Bequette 1997).

Gaudin painted a picture of a city on or beneath the ocean where solar, wind and wave power are harnessed as renewable energy sources, where seawater is desalinated and marine culture sustains the community. And while at the time Gaudin's vision may have come across as merely a resurgence of the longstanding quest to wrestle territory from the ocean, proposals such as Richard Dziewolski's *Marinarium*, a floating city with breakwaters, houses, two ports and a private beach along with facilities to match Gaudin's vision; and the underwater resort *Aquapolis* off the town of Mykonos, Greece, featured well resolved proposals that, while largely restricted to low-to-medium density eco-resort towns, were taking the first step toward realizing the marine utopias of the 1970s.

1.3   Advantages posed by ocean space colonization

Colonization of adjacent waters avails itself as a solution to population pressures and land scarcity. The reclaimed *Fontvieille* in Monaco accounts for 12% of Monaco's 2 sq. km. Likewise, Singapore has reclaimed about 100 sq. km amounting to 17% of its original land area and plans to reclaim more land are still underway. Begun in 1927, the *polders* of The Netherlands were reclaimed systematically from the *Zuiderzee* and when completed in 1967, added a total of 1,650 km of fertile agricultural land with an extensive fresh water supply.

But beyond relieving existing heavily used coastal land, offshore colonization also allows for the accommodation of socially sensitive facilities such as sewerage treatment and waste disposal plants, nuclear power plants, oil storage bases such as the *Shirashima Oil Storage Base* and airports such as the *Kansai International Airport* in Japan. These facilities can be housed on artificial islands in readily available ocean space located away from neighborhoods, yet near enough for easy access to the city. In addition, access to floating structures may be controlled thus enabling facilities that require a higher level of security to be housed offshore.

Military bases may be accommodated in the ocean as well. The United States Office of Naval Research has been engaged in studies on the technical feasibility and costs of building mobile offshore bases or sea bases. These bases will comprise self-propelled, modular, floating platforms that assemble (in any sea state) to provide logistic support for US military operations where fixed bases are not available. Security, again, may be easily provided for due to its isolated condition in the ocean.

Furthermore, the creation of artificial islands also allows for an increased shoreline. Dubai's *Palm* project will increase her shoreline by 166%, producing an extra 120 km of valuable waterfront land parcels ready for exploitation of the surrounding marine environment by leisure industries.

The impetus for ocean space colonization is fuelled also by the development of alternative sources of energy and the exploitation of existing hydrocarbon reservoirs. Hydrocarbons continue to remain the dominant energy source of today with "twenty percent of the oil and gas production (taking) place in offshore areas" (Moan 2003). New reserves are constantly being sought for in deeper waters with the aid of emerging technologies developed to support this continued colonization. Figure 1.1 shows the sedimentary basins of the world where potential hydrocarbon provinces may be located.

Methane hydrates (or burning ice) give out less carbon dioxide during combustion than hydrocarbons. These hydrates form at depths of 300–500 m near or below the sea floor and present an alternative source of future energy (Narita 2003). In addition, the ocean also provides energy through the harvesting of wave action, currents, offshore wind forces, and thermal differentials. Whilst work on exploiting wave energy in the 1970s failed to deliver an economically viable supply of electricity from wave energy (Moan 2003), there has been considerable advancement in the field.

The proposed *Pelamis Wave Energy Plant* and *Point Absorbers*, buoyant devices that move up and down with the waves, transmitting the motions to a suitable generator, are a few examples of these emergent wave energy converters. Similarly, the *Oscillating Water Column* (OWC) is a mechanical device in which the up and down movement of waves entering a chamber pumps air backward and forward through a pneumatic turbine which then generates electricity (Fraenkel 2000). The availability of wave energy grows progressively as one harvests further offshore, thus ocean space colonization plays a key role in future developments in wave energy conversion. Figure 1.2 displays the average wave power obtainable in various parts of the world.

Offshore winds may be harnessed through floating wind farms and other offshore wind installations, whilst their submerged counterparts would tap marine current energy allowing energy conversion to occur simultaneously below and above the sea. Germany and Denmark are countries that are aggressively pursuing wind energy for clean and sustainable development.

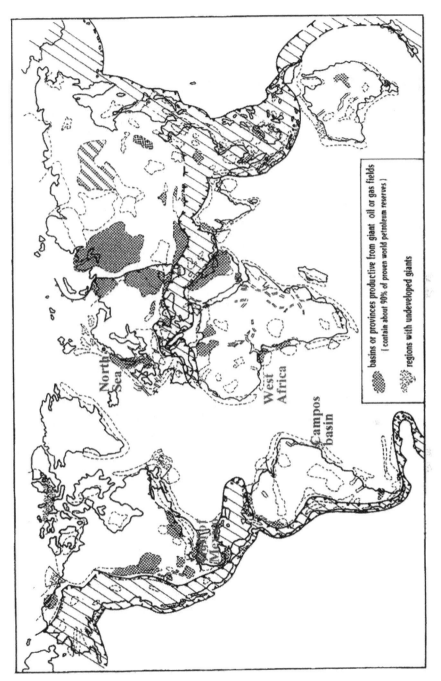

Figure 1.1 Sedimentary basins where potential hydrocarbon provinces may be located.

Source: Picture from Moan (2003).

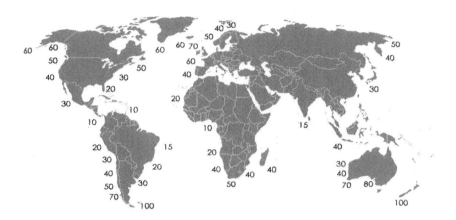

Figure 1.2 Worldwide availability of wave power (kW/m of wave front).

Source: Adapted from http://www.oceanpowertechnologies.com/advantages/ (accessed June 2004).

Temperature differentials may also be exploited through the use of Ocean Thermal Energy Conversion (OTEC) systems, which use the natural ocean thermal gradient to vaporize anhydrous ammonia, which in turn drives a turbine-generator in a closed loop, producing electricity (Vega 1999). The electricity may be used to convert water to hydrogen and oxygen by electrolysis. The hydrogen obtained when used in fuel cells is very environmentally friendly as the waste product from this way of obtaining energy is water.

The ocean also avails itself as a source of food. Moan (2003) posits that "the catch of wild fish in the world has reached an apparent limit of yield of about 100 million tonnes per year" and it appears that an increase in seafood production can only be achieved through marine culture. Countries such as the United States and Japan have built open-ocean ranches to enhance open-sea biological productivity and create sustainable fisheries. OTEC technology will also have an applicability here along with other upwelling mechanisms, employed to pump nutrient-rich water from the ocean depths onto the sea surface to attract fish and support marine life where little existed before.

Over 70% of world trade and 95% of the international transport are conducted by ships, with ports serving as an integral part of this transport system (Moan 2003). The integration between waterborne and land-based transport is crucial and with the new ability to transfer cargo at sea, ocean colonization may completely alter current economic balances and could bring ships once more into economic competition with airplanes. The scarcity of good harbors such as Athens' Piraeus, France's Cherbourg, Italy's Venice, the USA's New York, or Tokyo's Yokohama, can now be supplemented with port extensions into deeper waters and offshore floating ports. In time to come, Buckminster Fuller's dream of small cruising yachts sailing the world in "safe, one-day runs from one protected floating city's harbour to the next" (Kuromiya 1981) may finally be realized.

These arguments for the colonization of the sea are increasingly made pertinent through technological advances such as VLFS technology, which, if implemented, will allow for "a *Blue Revolution* that will offer advantages beyond mere agricultural commodities."

1.4   Evolution of VLFS technology

Large modular floating structures were first mooted by Armstrong in 1920s and were based on the US Navy *Sea Sled* of 1917 (Taylor 2003). These mobile offshore bases acted as "stepping stones" which allowed airplanes to skip over the Atlantic to refuel and rest. Armstrong's *Seadrome* of *c*.1940 was then amalgamated with existing technology employed in oil wells off Summerland, California and Baku, Russia by McDermott to install platforms in the Gulf of Mexico. Technological advancement on the floating concrete caissons developed in the North Sea by oil companies in the late 1970s and technology borrowed from the shipbuilding industry allowed the development of modular floating steel units that assemble to form a pontoon-type VLFS.

Suzuki (1997) defines VLFS not only as floating structures with large length dimensions, but also as having lengths larger than the characteristic length defined by the ratio of structural stiffness and buoyant spring stiffness. Owing to these structural dimensions, elastic responses in VLFSs are more dominant than their rigid-body motions.

Floating structures rely on the buoyancy force of the water in order to support themselves. In the broadest sense, soft seabed contact constructions where installations depend on a certain degree of buoyancy in order to reduce the reaction force on their supports are considered floating structures. However floating structures may be generally categorized as either pontoon-type or the semi-submersible type. The former is basically a simple box structure that features high stability with low manufacturing costs and easy maintenance. Pontoons are suitable for use only in calm water as they may roll or pitch being particularly prone to large wave-induced movement of the structure. They are often surrounded by a system of breakwaters to shelter the structure constructed in sea states with large wave heights. Shown in Figure 1.3 are the components of a floating pontoon structural system: the floating structure, the superstructure, the access-bridge, breakwater, and mooring facility. Floating structures may be moored with various systems depending on the requirements of the structure. For a greater restraint against horizontal movement, either the pier/quay wall method or the dolphin-frameguide mooring system may be adopted.

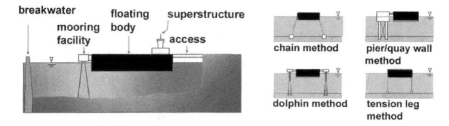

Figure 1.3 Left: Components of a floating pontoon structure. Right: Mooring methods.

The semi-submersible type finds its pioneers in the oil drilling rigs developed over the last 25 years in the North Sea and the Gulf of Mexico. These floating structures are designed to minimize the effect of waves on the structure while maintaining a constant buoyancy force. This particular type of floating structures is currently believed to be one of the few serviceable solutions for areas further away from shorelines where waves are larger and have been exploited as early as the 1970s. Kikutake's *Aquapolis* for the Oceanic Expo 75 in Okinawa, Japan, was modeled on technology adapted from the oil industry. An entry for the competition to redesign *Aquapolis* by Jacques Rougerie Architects in 1992 and present-day oil drilling rigs are examples of the continued relevance of this particular variety of floating structures. However, with the advent of the unique modular *WhisprWave Floating Breakwater*, an advanced version of Buckminster Fuller's floatable breakwater that assembles to attenuate wave forces (Fuller 1983), applications of pontoon-type VLFS in deeper waters are beginning to appear feasible in the near future.

1.5   Advantages of VLFS technology

The Floating Structures Association of Japan claims that floating structures may offer significant advantages over the traditional land reclamation technique for offshore colonization under certain conditions.

1.5.1   Economical for great water depths and soft seabed scenarios

As these structures depend on the inherent buoyancy force of the water, they are little affected by water depth or the nature of the seabed. Land reclamation becomes uneconomical at depths greater than 20 m and scenarios with soft seabed. Huge costs are incurred when the sand for reclamation has to be purchased. In Singapore, plans to increase land area to 820 sq. km through the reclaiming of another 140 sq. km (or 20.5% of its current land area) by 2010 requires 1.8 billion cu. m of sand. Sand is purchased at US$8.50 per cu. m from international brokers and translates to costs amounting to US$15.3 billion for sand alone (Guerin 2003). Furthermore, sand export and mining along with Singapore's aggressive land reclamation plans have caused political disputes between Singapore and Indonesia. An estimated 300 million cu. m was mined every year from the seabeds in Riau and Bangka-Belitung provinces adjacent to Singapore before Jakarta banned sand exports and mining in January 2002. This has led to the disappearance of numerous small islets in the province of Riau. Nipah Island, one of the 83 border islands which serve as points of reference for Indonesia's sea borders, is almost submerged. Indonesia has stressed that if the island sinks completely, the international boundary between Indonesia and Singapore will change in favor of the growing city-state (Guerin 2003).

1.5.2   Faster construction time

Reclamation work often requires consolidation time for land reclaimed to settle before construction may occur. This may be anywhere from two to five years if consolidation devices are used such as sand and jute drains. The period of consolidation increases by several years if the fill materials used are marine clay pumped from the seabed. Floating structures, on the other hand, may be occupied and utilized as soon as they are assembled. This significantly reduces the development period of the installation. The Floating Structures Association of Japan claims that construction time is further reduced as floating structures are modular and individual structural parts may be manufactured at plants in numerous locations with production of parts conducted parallel with on-site work. It took less than four months to construct the 1-km long Mega-Float, a floating test runway in the bay of Tokyo.

These claims, however, appear contextual in that costs of obtaining infill for land reclamation and that of labor and transporting individual structural parts vary from country to country. The proposed island cities off the coast of Tel Aviv, Israel will be built on reclaimed artificial islands, "a construction process that could take five to ten years" at a cost of US$2 billion ("Island cities planned off Israel's coast," *NewScientist.com News Service*, November 13, 2002). Should these islands be built in Japan using VLFS technology, the construction period would be significantly reduced with parallel production of structural parts and would be accomplished at a lower cost allowing the project to be monetized more rapidly. However, should this same technology be employed in a country without the same expertise and availability of manufacturing facilities as Japan, the cost and construction time required may well make the employment of VLFS technology uneconomical.

1.5.3   Ease of expansion and removal

The modular nature of floating structures also affords the advantages of mobility and flexibility that are not inherent in reclaimed land. Facilities can be removed if they become obsolete, towed and sunk as artificial reefs, or expanded and grouped with other floating structures as needed. Floating lodging facilities and plant barges have been built in Japan and towed to, and used in, the Middle East and other parts of the world. The sea space occupied by the floating structure may then be recovered.

Furthermore, the technology allows for the installation to be easily expanded and made to take any shape. For example, a dual loading for a floating container terminal requires a strip of water for the ship to be berthed. It would be very expensive for such a shape to be formed using land reclamation and sheet piling construction.

1.5.4   Environmentally friendly

Floating structures have less impact on the marine environment compared with land reclamation as they cause little underwater pollution and have negligible effect on tidal currents and thus avoid silting-up adjacent river mouths. Land reclamation interferes with the littoral flows of sand leading to the loss of natural flows in down-drift beaches. This generally creates a surplus condition on the up-drift side and a starvation condition on the down-drift beaches. The proposed island cities in Israel have been criticized by Philip Warburg, executive director of the Israel Union for Environmental Defence as "a barrier to the natural northward flow of sand (which would) in turn severely erode beaches and possibly cliffs for miles north of the (proposed islands)." Each of the artificial islands could potentially interrupt movement of between 50,000 and 200,000 cu. m of sand every year. This translates to annual costs of a million dollars to shift the built-up sand around each island (see for example "Island cities planned off Israel's coast," *NewScientist.com News Service*, November 13, 2002).

Malaysia has criticized Singapore's land reclamation works pointing out that the doubling of two islands (Pulau Tekong Besar and Pulau Tekong Kecil) in the Johor Strait would reduce flow areas and create stronger currents and tidal waves that would speed up sedimentation and block river discharges turning settlements along the rivers into flood-prone areas ("Flood threat from Singapore alleged," *The Straits Times*, March 21, 2002). Malaysia further alleged that reclamation works were affecting the water quality and depleting marine resources in the Johor Strait, in turn impacting the livelihoods of local fishermen ("Singapore reclamation works 'hurt Johor fishing'," *The Straits Times*, July 9, 2002).

Currently, the Floating Structures Association of Japan and the Shipbuilding Research Centre of Japan list, as part of their principal conditions for the adoption of floating structures, the requirement of having minimal environmental effects during and after construction, accomplished through a Maritime Environmental Impact Assessment. From initial tests, it has been found that the floating structure has little impact on marine environments 6 m below the surface. Impact on marine life can be further reduced through cut-outs in the floating structure to allow greater sunlight penetration.

1.5.5   Inherently base isolated

Floating structures are protected from seismic shocks since they are inherently base isolated. The floating *K-CAT terminal* in Kobe (Figure 1.4), a pier for high-speed shuttle boats linking the Kansai International Airport and Kobe Port Island suffered no damage in the Great Hanshin Earthquake of 1995, which measured 7.2 on the Richter scale. Also due to their base-isolation, they are ideal for emergency bases and can be towed to the disaster site. Japan has a number of floating rescue emergency bases such as the ones in Tokyo Bay, Osaka Bay and Ise Bay (see Figure 1.5).

In addition, floating structures respond well to tidal changes as the surface of the floating structure relative to the water surface is constant. This particular characteristic avails VLFS technology to applications that call for excellent closeness to the water such as marinas, docks, piers and other leisure facilities.

1.5.6   Reduced requirements for foundation

The greatest advantage offered by floating structures is that it simply harnesses the buoyancy force of water to support itself. This translates to a

Figure 1.4 The floating K-CAT terminal, Kobe, Japan.

Source: *Look at the Sea*. The Floating Structures Association of Japan.

Figure 1.5 Floating rescue emergency base moored at Tokyo Bay.
Source: *Look at the Sea.* The Floating Structures Association of Japan.

reduction in construction costs associated with providing adequate support to the installation. As such, VLFS technology is employed extensively in the field of bridge-building where all of the above advantages of floating structures are exploited. Norway's 845-m long Bergsoysund Bridge over a fjord depth of 320-m and 1,246-m long Nordhordland Bridge over a fjord depth of 500 m (Figure 1.6), and the United States' 2,013-m long Lacey V. Murrow Bridge and the Third Washington Bridge over Lake Washington, Seattle (see Figure 1.7) and Japan's Yumemai floating steel arch bridge (see Figure 1.8) are but a few examples. It is worth noting that land based structures need massive and immovable foundations while VLFS require much less massive foundation system. In fact the mooring system of the VLFS allows free vertical movement of the structure to follow the changing sea level and allows slight movements in the horizontal direction.

1.5.7 Availability of interior spaces

In order to achieve the buoyancy, very large floating structures have many watertight compartments. These interior spaces may be used for car parks, offices and storage rooms. For example, the floating breakwater in Monaco, which comprises of a large double hulled measuring 353-m long, 44-m wide and 24-m high precast structure, houses a 380-lot car park and a dry dock for recreational craft (*Concrete Technology*, Vol. 5, No. 2, 2006, p. 61).

Figure 1.6 Nordhordland Bridge built in 1994 at Salhus, Norway. Photo courtesy of E. Watanabe.

Figure 1.7 Lacey V. Murrow Bridge and Third Washington Bridge at Seattle, USA. Photo courtesy of E. Watanabe.

Figure 1.8 Yumemai floating bridge built in 2000 at Osaka, Japan.

1.6 Current applications of VLFS technology

VLFS technology has been employed in numerous ocean environment structures. These include installations that take advantage of its ability to locate socially sensitive facilities away from existing urban development, such as floating storage bases like the Kamigoto Oil Storage Base and the Shirashima Oil Storage Facility (see Figure 1.9) in Japan; and sewage treatment plants like the facilities in Shizuoka Prefecture, Japan. The floating storage modules are huge. For example, in the Shirashima Oil Storage Facilitiy, one floating storage module measures 397 m × 82 m × 25.4 m and contains 700,000 kl of fuel that Japan consumes per day.

The Floating Structures Association is also proposing floating airports that will be cheaper and faster to construct than current offshore airports which sit on reclaimed land. Moreover, the flatness of land and their distance from urban developments make these floating airports attractive. The floating airport will have breakwaters installed along the perimeter of the site in optimum combination with mooring *dolphins* that anchor the floating structure, so as to minimize adverse wave effects. A 1-km long floating test runway was constructed in Tokyo bay to study the feasibility of floating airports (Figure 1.10, Suzuki 2005). This very large floating structure was called the Mega-Float by the Japanese and it was entered into the Guinness Book of Records as the world's largest man-made island. The city of Vancouver in Canada and the city of New York in USA have floating heliports (see Figure 1.11).

Other installations capitalize on the mobility afforded to floating structures. These include Sten Sjostrand's 200-room floating *Four Seasons Great Barrier Reef Hotel* that was located in Townsville, Australia. The hotel was initially built in Singapore for US$22 million and subsequently floated to Australia where it was anchored at two ends to the ocean floor. In the event of a cyclone, one of the ends would be disconnected after the hotel

Figure 1.9 Kamigoto and Shirashima floating oil storage bases, Japan.

Source: *Look at the Sea*. The Floating Structures Association of Japan.

had been evacuated, allowing the hotel to spin around in the water, riding out the storm. The hotel was eventually floated to Ho-Chi-Minh, Vietnam after the venture at Townsville went bankrupt. Now the hotel is moored in North Korea. Similar applications include floating lodging facilities, pulp plant barges, and seawater desalination plant barges that constantly operate in different locations.

Russia is planning to construct the world's first floating nuclear power station. Such mini power plants (1/150th of the power produced by a standard Russian nuclear power plant) would provide electricity and head to regions with undeveloped infrastructure. The mobile nature of the floating nuclear power plants would purportedly allow them to be moved to areas struck by natural disasters or other emergencies. The plants could also be used for desalination of sea water.

Figure 1.10 Mega-float (1,000 m × 60–120 m × 3 m) in Tokyo Bay. Photo courtesy of E. Watanabe.

Figure 1.11 Floating heliport in Vancouver, Canada.

Source: *Look at the Sea*. The Floating Structures Association of Japan.

Structures such as the *K-Cat terminal* mentioned earlier, the *Yokohama Passenger Terminal*, mooring and ferry piers in Hiroshima (see Figure 1.12), in Vancouver and the floating berth for a container terminal at Valdez, Alaska maintain the desirable constant relative distance between the floating structure and water level. Likewise, floating bridges such as the *West India Quay Footbridge* in London (Figure 1.13) and the *Nordhordland Bridge* in

Figure 1.12 Ujina's floating concrete pier for ferries and boats in Hiroshima, Japan.

Figure 1.13 West India Quay footbridge, London, UK. Photo courtesy of T. Utsunomiya.

Norway take advantage of this closeness to the water as well as the buoyancy force to span large distances with minimal support.

Singapore has constructed its first large floating platform (120 m × 83 m × 1.2 m) in the Marina Bay in year 2007 (see Figure 1.14). The platform comprises 15 individual steel pontoons joined together by bolted connections. This multipurpose floating platform will be used for Singapore's National Day Parade and other events such as concerts and sports competition.

Figure 1.14 Mega floating platform, Marina Bay, Singapore.

It is able to take the weight of 9,000 people, 200t of stage props and three 30-t vehicles (*The Straits Times*, October 17, 2005 and January 12, 2007). The floating platform is held in position by six perimeter mooring piles and is connected to the embankment by three linkways. An interesting feature of this platform is that it is designed so that it can be dismantled readily and towed away when the sea space is needed as well as the versatility it possesses to be reconfigured into different platform shapes if desired.

Floating island cities, either moored to coastal host cities or free-floating cities in international waters, appear to be the next progression for VLFS technology. For example, the Japan Society of Steel Construction proposed a floating city in one of the foci of Osaka Bay as shown in Figure 1.15. However, for this technology to develop beyond its current applications in bridge-building and eco-resorts to realize floating cities, the "long-term dream of the Floating Structures Association of Japan," depends on a number of factors beyond its technological serviceability. These include issues of economic viability, legality surrounding the colonization of ocean space, and the psychology of gated and deed restricted communities, utopia and micro-nationalism that affect public sentiment toward these schemes.

1.7   Concluding remarks

Ocean space will progressively be colonized by mankind in search of space, energy and food. However, it is important to utilize the ocean in an environmentally friendly and sustainable way, otherwise we could well destroy the remaining beautiful resources that we have. Innovative technologies that

Figure 1.15 Proposed floating city in Osaka Bay by Japan Society of Steel Construction.
Source: Japan Society of Steel Construction.

exert a light urban footprint on the environment, such as very large floating structures, are extent and available for all sorts of applications. It is this belief that led the authors to promote such technology to countries that are hungry for more space and to make use of their coastal and sea spaces.

References

Bequette, F. (1997) "Waterworlds," *UNESCO Courier*, Vol. 49, No. 6, pp. 45–47.

Davidson, N., Laffoley, D., Doody, J., Way, L., Gordon, J., Key, R., Drake, C., Pienkowski, M., Mitchell, R., and Duff, K. (1991) *Nature Conservation and Estuaries in Great Britain*. Peterborough, UK: Nature Conservancy Council, p. 422.

Fraenkel, P. (2000) "Renewable Option 5: Marine Power—Harvesting the Power of the Sea." *People and Planet.net* http://www.peopleandplanet.net/doc.php?id=452 (accessed June 2004).

Fuller, R. B. (1983) *The Patented Works of R. Buckminster Fuller*. New York: St. Martin's Press, pp. 269–273.

Fuse, T. (2003) "Legal Framework for the Use of Oceans—Considerations from International Law of the Sea," Proceedings of International Symposium on *Ocean Space Utilization Technology*, Tokyo, Japan, p. 3.

Guerin, B. (2003) "The Shifting Sands of Time—and Singapore," *Asia Times*, July 31.

Healy, M. (1995) "European Coastal Management: An Introduction," in Healy, M. and Doody, J. (eds), *Directions in European Coastal Management*. Tresaith, Cardigan, UK: Samara Publishing, pp. 1–6.

Healy, M. and Hickey, K. (2002) "Historic Land Reclamation in the Intertidal Wetlands of the Shannon Estuary, Western Ireland," *Journal of Coastal Research, Special Issue* 36, pp. 365–373.

Houston, M. (2004) "This Family is Living the New Australian dream...," *The Sunday Age Magazine*, April 18, pp. 16–19.

Kuromiya, K. (1981) *Critical Path*. New York: St Martin's Press. p. 331.

Matsuda, F., Sakou, T., Takahashi, M., Szyper, J., Vadus, J., and Takahashi, P. (1999) "U.S.–Japan Advances in Development of Open-Ocean Ranching," *US–Japan Cooperative Program in Natural Resources 23rd Meeting*. http://www.dt.navy.mil/ip/mfp/index.html (accessed June 2004).

McClung de Tapia, E. and Sugiura, Y. (2002) "Prehispanic Life in a Man-made Island Habitat in Chignahuapan Marsh, Santa Cruz Atizapan, State of México, México," *Foundation for the Advancement of Mesoamerican Studies, Inc.* http://www.famsi.org/reports/98024/index.html (accessed April 2004).

Moan, T. (2003) *Marine Structures for the Future*. CORE Report No. 2003-01. Singapore: National University of Singapore, p. 29.

Narita, H. (2003) "Introduction of MH21 (Research Consortium for Methane Hydrate Resources in Japan) and Current Topics in Production Method and Modeling of Methane Hydrate," *Proceedings of the Fifth ISOPE Ocean Mining Symposium*, Tsukuba, Japan.

Suzuki, H. (2005) "Overview of Megafloat: Concept, Design Criteria, Analysis and Design," *Marine Structures*, Vol. 18, pp. 111–132.

Suzuki, H., Yasuzawa, Y., Fujikubo, M., Okada, S., Endo, H., Hattori, Y., Okada, H., Watanabe, Y., Morikawa, M., Ozaki, M., Minoura, M., Manabe, H., Iwata, S., and Sugimoto, H. (1997) "Structural Response and Design of Large Scale Floating Structure," *Proceedings of OMAE*, Yokohoma, Japan, Vol. VI, pp. 131–137.

Taylor, R. (2003) "Mobile Offshore Base Project Summary and Technology Spin-offs," Proceedings of International Symposium on *Ocean Space Utilization Technology*, Tokyo, Japan, p. 30.

Thomsen, C.W. (1994) *Visionary Architecture: From Babylon to Virtual Reality*, New York: Prestel-Verlag, p. 35.

Vega, L. (1999) "Ocean Thermal Energy Conversion (OTEC)," *Strategic Industries Division, State of Hawaii*. http://www.hawaii.gov/dbedt/ert/otec/index.html (accessed June 2004).

Chapter 2

Wave phenomenon and properties

Seiya Yamashita

2.1 Small amplitude wave theory

Water wave may be defined as the deformation of free water surface. The wave form propagates far away while the water particle excited by the wave does not propagate along with the wave form. Figure 2.1 indicates the general form of the train of the idealized regular wave. The wave height is defined as the vertical distance from the trough to the crest of the wave while the wave length is the horizontal distance between successive wave crests.

The motion of the fluid with the free water surface satisfies three kinds of boundary condition. The first kind is the dynamic condition which indicates that the pressure on the free water surface is equivalent to the atmospheric pressure. The second kind is the kinematic condition which means that the water particle does not move across the free water surface. The third kind is the seabed condition. Based on the assumptions that water is inviscid,

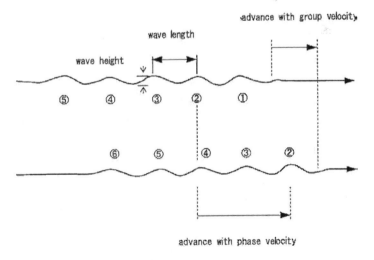

Figure 2.1 Wave propagation.

incompressible and its flow is irrotational, the motion of water particles can be characterized by a quantity known as the velocity potential. The velocity potential that satisfies the Laplace equation in the fluid domain is defined as a function whose derivatives yield the velocity component of the fluid. For small-amplitude waves where the wave height is assumed to be very much smaller than the wave length, the free-surface boundary conditions are expressed as

$$\zeta = -\frac{1}{g}\frac{\partial \Phi}{\partial t}, \quad \text{on } z = 0 \quad \text{dynamic condition} \tag{2.1}$$

$$\frac{\partial \Phi}{\partial z} = \frac{\partial \zeta}{\partial t}, \quad \text{on } z = 0 \quad \text{kinematic condition} \tag{2.2}$$

and the sea bed condition is given by

$$\frac{\partial \Phi}{\partial z} = 0, \quad \text{on } z = -h \tag{2.3}$$

where z is the vertical coordinate measured from the origin which is taken at the still-water, level h the water depth, g the gravitational acceleration, $\Phi(x, y, z; t)$ the velocity potential and $\zeta(x, y; t)$ the surface elevation. Equations (2.1) and (2.2) may be combined to give

$$\frac{\partial^2 \Phi}{\partial t^2} + g\frac{\partial \Phi}{\partial z} = 0, \quad \text{on } z = 0 \tag{2.4}$$

The velocity potential which satisfies the boundary conditions (2.3) and (2.4) is given by

$$\Phi(x, y, z; t) = \text{Re}\left[\frac{ig\zeta_A}{\omega}\frac{\cosh k(z + h)}{\cosh kh}e^{ik(x\cos\alpha + y\sin\alpha) + i\omega t}\right] \tag{2.5}$$

where ζ_A is the amplitude, ω the circular frequency, k the wave number and α the incident angle of the waves with respect to the positive x-axis as shown in Figure 2.2. By substituting Eq. (2.5) into Eq. (2.4), one obtains

$$\frac{\omega^2}{g} = k\tanh kh \tag{2.6}$$

Equation (2.6) is known as the linear dispersion relation.

The surface elevation, or the wave train, obtained from Eqs (2.1) and (2.5) may be expressed as

$$\zeta(x, y; t) = \text{Re}\left[\zeta_A e^{ik(x\cos\alpha + y\sin\alpha) + i\omega t}\right] \tag{2.7}$$

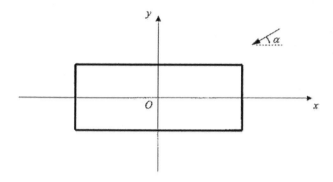

Figure 2.2 Definition of incident angle of wave.

When discussing waves that propagate only along the negative x-axis, the expression (2.7) may be simply written as

$$\zeta(x;t) = \mathrm{Re}\left[\zeta_A e^{i(kx+\omega t)}\right] = \zeta_A \cos(kx + \omega t) \tag{2.8}$$

The velocity of the wave is the traveling speed of the wave profile which is called the phase velocity. The expression for the velocity of the wave profile is given by

$$c = \frac{\omega}{k} = \sqrt{\frac{g}{k} \tanh kh} \tag{2.9}$$

and it is derived from Eq. (2.6). This indicates the wave velocity depends on just the wave number. On the other hand, the velocity of the harmonic wave can be defined by

$$c = \frac{\lambda}{T} \tag{2.10}$$

where λ is the wave length and T the wave period. The combination of Eqs (2.9) and (2.10) yields

$$k = \frac{2\pi}{\lambda} \tag{2.11}$$

which gives the relation between the wave number and the wave length.

There are two kinds of velocity for the wave train, namely, the phase velocity as mentioned earlier and the group velocity. A group of the wave train in Figure 2.1 advances with a group velocity whereas individual waves in the group move along with the phase velocity. Individual waves successively reach the leading wave in the wave train, because the phase velocity is greater than group velocity. The magnitude of the group velocity is half of the phase velocity for the dispersive wave in deep water. The group velocity for a non-dispersive wave such as swell is equal to the phase velocity.

Wave velocity depends on the magnitude of the wave height, and increases with increasing wave heights. The wave velocity derived from the small-amplitude wave theory is independent of the wave height. The wave velocity, however, is a good approximation for the velocity of a large amplitude wave.

Water depth may be regarded as infinite if the water depth is greater than half of the wave length (i.e., $h > \lambda/2$). As $h \to \infty$ in Eqs (2.5) and (2.6), the velocity potential for the incident wave and the dispersion relation are expressed respectively by

$$\Phi(x, y, z; t) = \text{Re}\left[\frac{ig\zeta_A}{\omega}e^{Kz+iK(x\cos\alpha+y\sin\alpha)+i\omega t}\right] \tag{2.12}$$

$$\frac{\omega^2}{g} = K \tag{2.13}$$

In very shallow water (roughly $h < \lambda/20$), the wave is usually referred as a long wave and the dispersion relation becomes

$$\frac{\omega^2}{g} = k^2 h \tag{2.14}$$

as $h \to 0$ in (2.6). This relation gives

$$c = \frac{\omega}{k} = \sqrt{gh} \tag{2.15}$$

which shows that the wave velocity is independent of the wave length and the long wave is not dispersive.

Table 2.1 summarizes the expression for the fundamental properties of small-amplitude waves. The range of water depth in Table 2.1 is divided into deep water, intermediate depth and shallow water.

2.2   Pressures of fluid motion

When the wave train travels on the free water surface, water particles are excited in motion. The path of a particular particle in a complete cycle takes a circular path having a radius which is equivalent to the wave amplitude at the surface in the case of deep water. The radius, however, decreases with depth from the surface in proportion to e^{Kz} ($z < 0$). In the case of intermediate water depth, the particles move in ellipses while in shallow water, the ellipses have a constant horizontal distance between foci as shown

Table 2.1 Fundamental properties of waves

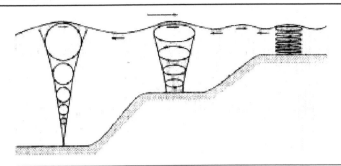

	Deep water	Intermediate depth	Shallow water
	$h > \dfrac{\lambda}{2}$		$h < \dfrac{\lambda}{20}$
Dispersion relation	$\dfrac{\omega^2}{g} = K$	$\dfrac{\omega^2}{g} = k\tanh kh$	$\dfrac{\omega^2}{g} = k^2 h$
Phase velocity	$c = \dfrac{\omega}{K} = \dfrac{g}{2\pi}T \cong 1.56T$	$c = \dfrac{\omega}{k} = \sqrt{\dfrac{g}{k}\tanh kh}$	$c = \dfrac{\omega}{k} = \sqrt{gh}$
Wave length	$\lambda = cT = \dfrac{g}{2\pi}T^2 \cong 1.56T^2$	$\lambda = cT = \dfrac{2\pi}{k}$	$\lambda = cT = T\sqrt{gh}$
Group velocity	$c_g = \dfrac{c}{2}$	$c_g = \dfrac{c}{2}\left(1 + \dfrac{2kh}{\sinh 2kh}\right)$	$c_g = c$
Wave energy	$\bar{E} = \tfrac{1}{2}\rho g \zeta_A^2$	$\bar{E} = \tfrac{1}{2}\rho g \zeta_A^2$	$\bar{E} = \tfrac{1}{2}\rho g \zeta_A^2$
Wave power	$W = \bar{E}c_g \cong 0.98H^2 T$	$W = \bar{E}c_g$	$W = \bar{E}c_g \cong 3.93H^2\sqrt{h}$

Notes

$H = 2\zeta_A$ is the wave height; $\bar{E}$ is the wave energy per unit area of the free water surface (Nm/m²); W is the wave power (energy flux) per unit width in y-direction (kW/m).

in Table 2.1. At the seabed, the particles oscillate back and forth on a straight line for the cases of intermediate depth and shallow water.

The expressions for the horizontal and vertical components of the velocity of water particles are given by

$$u = \frac{\partial \Phi}{\partial x} = \text{Re}\left[-\omega\zeta_A \cos\alpha \frac{\cosh k(z+h)}{\sinh kh}e^{ik(x\cos\alpha+y\sin\alpha)+i\omega t}\right] \qquad (2.16)$$

$$w = \frac{\partial \Phi}{\partial z} = \text{Re}\left[i\omega\zeta_A \frac{\sinh k(z+h)}{\sinh kh}e^{ik(x\cos\alpha+y\sin\alpha)+i\omega t}\right] \qquad (2.17)$$

in view of the dispersion relation (2.6). It is noted that under the wave crest, the horizontal velocity of the particle is in the wave propagation direction, while under the wave trough the velocity is opposite to the wave propagation direction. The amplitude of horizontal velocity decreases with water depth in proportion to $\cosh k(z + h)$, while the amplitude of vertical velocity decreases in proportion to $\sinh k(z + h)$ and diminishes at the seabed ($z = -h$). As mentioned above, the water particles oscillate horizontally at the sea bed.

The motion of the water particle of sinusoidal waves yields the hydrodynamic pressure which varies harmonically in time. The total pressure including the hydrostatic pressure follows from the Bernoulli equation. One can write

$$P(x, y, z; t) = -\rho \frac{\partial \Phi}{\partial t} - \frac{1}{2}\rho(u^2 + v^2 + w^2) - \rho g z \tag{2.18}$$

where $-\rho g z$ is the hydrostatic pressure. This equation is valid for unsteady, irrotational and inviscid fluid motion. For small-amplitude waves, the second-order terms in the wave amplitude can be neglected. Thus, the pressure may be approximated as

$$P(x, y, z; t) = -\rho \frac{\partial \Phi}{\partial t} = \mathrm{Re}\left[\rho g \zeta_A \frac{\cosh k(z + h)}{\cosh kh} e^{ik(x \cos \alpha + y \sin \alpha) + i\omega t} \right] \tag{2.19}$$

by disregarding the hydrostatic pressure which is independent of the motion of water particle. The pressure due to the wave motion is proportional to the wave amplitude and it reduces with respect to the water depth according to $\cosh k(z+h)$. By comparing Eqs (2.7) and (2.19), one gets a positive dynamic pressure under a wave crest and a negative dynamic pressure under a wave trough.

2.3   Wave energy

The energy of a train of sinusoidal waves is composed of kinetic energy and potential energy. The kinetic energy is associated with the orbital motion of water particles, while the potential energy is resulting from the change of water level in wave crests and troughs. If one considers a water column having a unit area of the free water surface as shown in Figure 2.3, the kinetic and potential energies of the small portion of the column whose height is dz, is given by

$$dE = \left\{ \tfrac{1}{2}\rho(u^2 + w^2) + \rho g(z + h) \right\} dz \tag{2.20}$$

Here, it is assumed that the waves propagate along the x-axis.

The wave energy is, in general, defined by the energy averaged with respect to time over one cycle of the wave motion. The total mean energy which is

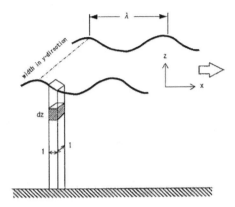

Figure 2.3 Water column in wave.

the average energy per unit area of the free water surface, can be obtained by the integration of the energy (2.20) as follows:

$$\overline{E} = \frac{1}{T} \int_0^T \left\{ \int_{-h}^0 \frac{1}{2}\rho(u^2 + w^2)dz + \int_{-h}^\zeta \rho g(z + h)dz \right\} dt$$
$$= \frac{1}{4}\rho g \zeta_A^2 + \frac{1}{4}\rho g \zeta_A^2$$
$$= \frac{1}{2}\rho g \zeta_A^2 \tag{2.21}$$

by exploiting Eqs (2.16), (2.17) and (2.8). The bar denotes the time average. This means that for small-amplitude waves, the kinetic energy is identical to the potential energy and the wave energy (or the energy density) is proportional to the square of the wave height.

The energy flux of the wave occurs across the vertical control surface. The mean rate of the energy flux per unit width in the y-direction across the vertical surface $x =$ constant is given by

$$\overline{\frac{dE}{dt}} = \frac{1}{T} \int_0^T \int_{-h}^0 \left(-\rho \frac{\partial \Phi}{\partial t} \right) \left(-\frac{\partial \Phi}{\partial x} \right) dzdt \tag{2.22}$$

The substitution of the velocity potential (2.5) into Eq. (2.22) yields

$$\overline{\frac{dE}{dt}} = \overline{E}\frac{1}{2}\frac{\omega}{k}\left(1 + \frac{2kh}{\sinh 2kh} \right) = \overline{E}c_g \tag{2.23}$$

where c_g is the group velocity of the wave train. This means that the energy flux of the wave, which is sometimes referred to as the wave power, is the product of the wave energy and the group velocity. The energy flux of the wave is constant for any depth of water. Therefore, the wave height is inversely proportional to the square root of the group velocity.

2.4   Descriptions of irregular waves

The outstanding characteristic of the ocean wave is its irregularity. However, the complete randomness of the wave necessitates a statistical description of the wave. The wave height which varies from crest to crest in the ocean is specified by the average apparent height, or the significant wave height defined as the average of a third of the highest wave heights recorded. Similarly, the apparent average period, which is the time between successive crests or successive zero-up-crossings of the mean line, is used for characterizing the wave. The zero-up-cross and crest-to-crest methods for determining the wave height and period from the record of irregular waves are illustrated in Figure 2.4.

In the linear theory, the irregular waves can be represented by the sum of a large number of small-amplitude regular waves having different amplitudes, lengths and directions. This concept of superposition is very important and powerful for the description of the ocean waves. The wave elevation at a fixed position in the ocean can be written as

$$\zeta(t) = \sum_{i=1}^{N} a_i \sin(2\pi f_i t + \varepsilon_i) \tag{2.24}$$

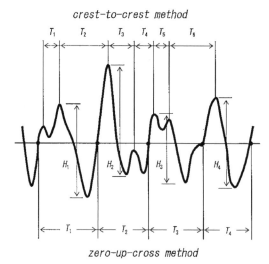

Figure 2.4 Irregular wave. Top: Crest-to-crest method. Bottom: Zero-up-cross method.

where a_i, f_i and ε_i denote respectively the amplitude, the frequency and the random phase of each component of the wave. The root mean square σ of the wave elevation has an important role on the statistical description of the ocean waves. The root mean square is defined by

$$\sigma^2 = \frac{1}{N}\sum_{i=1}^{N}\zeta_i^2 \xrightarrow[N\to\infty]{} \frac{1}{T}\int_0^T \zeta^2(t)\,dt \quad (T\to\infty) \tag{2.25}$$

and the substitution of Eq. (2.24) into Eq. (2.25) gives

$$\sigma^2 = \frac{1}{2}\sum_{i=1}^{N} a_i^2 \tag{2.26}$$

This shows that the mean square of the irregular wave is proportional to the sum of the energy of the component wave. The relationships between significant wave height $H_{1/3}$ and the root mean square σ are discussed later.

2.5 Wave spectrum

The wave spectrum represents the distribution of the wave energy versus frequency of component waves comprising irregular waves. Referring to Figure 2.5, any increment of area under the graph is proportional to the wave energy which is given by Eq. (2.21), that is

$$S(f_i)\Delta f = \frac{1}{2}a_i^2 \tag{2.27}$$

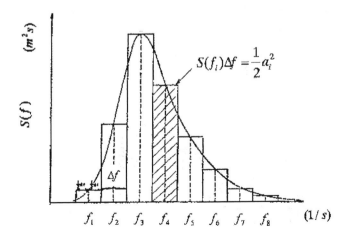

Figure 2.5 Illustration of wave spectrum.

where $S(f)$ is the wave spectrum is (sometimes referred to as the spectral density) and Δf means the band of frequency that is a constant difference between successive frequencies.

The total energy of the irregular wave is the sum of all the component energies. It can be expressed as

$$\frac{1}{2}\sum_{i=1}^{N} a_i^2 = \sum_{i=1}^{N} S(f_i)\Delta f \xrightarrow[N\to\infty]{} \int_0^\infty S(f)\,df \tag{2.28}$$

More precisely, the sum is proportional to the wave energy per unit surface area in irregular waves. It is found by substitution of Eq. (2.26) into Eq. (2.28) that the mean square value is equal to the area under the wave spectrum, that is,

$$\sigma^2 = \int_0^\infty S(f)\,df \tag{2.29}$$

This indicates that if the wave spectrum is known, one can determine the root mean square value of the recorded waves.

The wave spectrum can be estimated from the measurements of the ocean waves. Several wave spectra for fully developed seas are recommended for evaluating $S(f)$ based on the measurements. They are generally written in the form of

$$S(f) = Af^{-5}\exp[-Bf^{-4}] \tag{2.30}$$

where A and B are constant coefficients which are determined by the significant wave height and the average wave period. The expressions of A and B are given in Table 2.2 for typical wave spectra, such as the Bretschneider-Mitsuyasu spectrum, the ISSC (International Ship and Offshore Structure Congress) spectrum and the JONSWAP (Joint North Sea Wave Project) spectrum. In the table, $T_{1/3}$, T_{01} and T_P are respectively the significant period, the mean period and the period corresponding to the peak frequency of the wave spectrum.

Table 2.2 Constant coefficients for wave spectra

Spectrum	Wave period	A	B
Bretschneider–Mitsuyasu	$T_{1/3}$	$0.257H_{1/3}^2 T_{1/3}^{-4}$	$1.03T_{1/3}^{-4}$
ISSC	T_{01}	$0.111H_{1/3}^2 T_{01}^{-4}$	$0.44T_{01}^{-4}$
JONSWAP	T_P	$\alpha\gamma^\beta H_{1/3}^2 T_P^{-4}$	$1.25T_P^{-4}$
	$(=1.05T_{1/3})$		

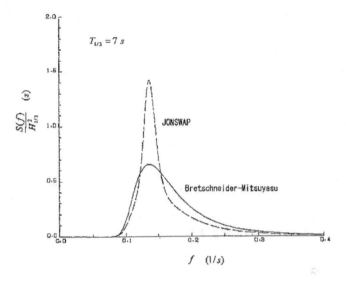

Figure 2.6 Examples of wave spectra.

Comparisons of the Bretschneider–Mitsuyasu spectrum and the JONSWAP spectrum are illustrated in Figure 2.6. If the wave spectrum is written by using the circular frequency $\omega \ (=2\pi f)$, the following formula can be used for $S(\omega)$

$$S(\omega) = \frac{1}{2\pi} S(f) \tag{2.31}$$

The short-crestedness of the waves should be considered for a more complete representation of the seaway. A short-crested wave can be characterized by a two-dimensional wave spectrum, which is often referred to as directional spectrum which can be approximated by

$$S(f,\theta) = S(f)G(\theta) \tag{2.32}$$

where $G(\theta)$ means a function of the angle of wave propagation direction. An example of $G(\theta)$ is given by

$$G(\theta) = \frac{2}{\pi}\cos^2\theta, \quad -\frac{\pi}{2} \le \theta \le \frac{\pi}{2} \tag{2.33}$$

Here, $\theta = 0°$ corresponds to the main direction of wave incidence.

2.6   Stochastic properties

A good approximation to the probability density function can be derived for the wave height from the so-called Rayleigh distribution. The mathematical expression of the distribution is given by

$$p(H) = \frac{H}{4\sigma^2} \exp\left[-\frac{H^2}{8\sigma^2}\right] \tag{2.34}$$

where H is the height of individual waves and σ the root mean square defined by Eq. (2.25). By assuming that the waves are characterized by a narrowbanded spectrum in which the components of the wave frequencies are concentrated over a narrow range of the frequency, the following relation between the root mean square of the wave height H_{rms} and the root mean square value σ exists:

$$H_{rms} = 2\sqrt{2}\sigma \tag{2.35}$$

The assumption of the Rayleigh distribution also gives the relation between the significant wave height and the root mean square of the wave height. Consequently, the following useful relation is obtained:

$$H_{1/3} = \frac{H_{rms}}{0.706} = 4.0\sigma \tag{2.36}$$

In the ocean, the significant wave height and average wave period vary in long years. A long-term description of the sea needs the data for the variation of the significant wave height and average wave period. That is to say, the joint frequency of the significant wave height and average wave period as illustrated in Table 2.3 is necessary for the long-term prediction of the sea. For instance, the probability of occurrence p_5 which represents the probability of the wave height between 2 and 2.5 m in Table 2.3 is

$$p_5 = \frac{212 + 1,360 + 731 + 114 + 7}{15,071} = 0.161 \tag{2.37}$$

Considering that the wave height follows the Rayleigh distribution, the long-term probability can be obtained as

$$P(H) = \sum_{i=1}^{M} \exp\left[-\frac{2H^2}{\left(H_{1/3}^{(i)}\right)^2}\right] p_i \tag{2.38}$$

Table 2.3 Joint frequency of significant wave height and average wave period

$H_{1/3}$ (m)	T (s)					
	4.5	6.5	8.5	10.5	12.5	14.5
0.0–0.5	373	12				
0.5–1.0	1,960	184	29	3		
1.0–1.5	1,990	948	134	26	6	
1.5–2.0	1,079	1,854	358	69	13	
2.0–2.5	212	1,360	731	114	7	
2.5–3.0	38	448	690	230	43	1
3.0–3.5	9	228	448	182	53	6
3.5–4.0	5	55	249	77	42	6
4.0–4.5	3	50	209	87	20	19
4.5–5.0		9	68	38	10	7
5.0–5.5		9	44	44	8	3
5.5–6.0		4	4	17	14	1
6.0–6.5			27	41	11	1
6.5–7.0				3	2	
7.0–7.5			4	8	9	
7.5–8.0			1	2	2	
8.0–8.5			2	3	4	1
8.5–9.0						
9.0–9.5				1	4	1
9.5–10.0						
10.0–10.5			1			
10.5–11.0						
11.0–11.5					2	1

Note
Total sum of frequency is 15,071.

where $P(H)$ is the probability that the individual wave height exceeds H. If an average period of wave during 100 years in a sea is assumed to be 7 s, it may be given by

$$P(H) = \frac{1}{N} = \frac{7}{60 \times 60 \times 24 \times 365 \times 100} \cong 10^{-8.7} \qquad (2.39)$$

from the total frequencies of the wave N during 100 years. Relationship (2.38) gives the wave height H corresponding to this probability of occurrence.

2.7   Concluding remarks

Knowledge about the ocean waves is important for the design of a VLFS in both operating and survival conditions. Exciting wave forces (that trigger the elastic response of the VLFS), mean drift forces and slowly varying drift forces (that are required for the design of mooring system of the VLFS) are directly influenced by the wave height and wave period in the ocean.

The short-term sea states are required for investigating the motion characteristics of the VLFS in the operating condition in which the vertical motion is a limiting factor. Wave spectrum described in Section 2.5 provides important information for the simulation of VLFS motions in the short-term sea states.

In general, an extreme wave height (e.g. the wave height of a 100-year wave) is used for the structural design of a VLFS in survival conditions for which the breaking strength of the VLFS should be investigated. The extreme wave is considered as an individual wave in the long-term sea states. Therefore the wave height has to be predicted in a statistical manner by using the long-term wave data such as those given in Table 2.3 of Section 2.6.

Chapter 3

Hydroelastic analysis of VLFS

Tomoaki Utsunomiya

3.1 Introduction

A typical very large floating structure (VLFS) has large horizontal dimensions ranging from several hundred meters to several kilometers. On the other hand, the depth of the VLFS is only several meters. For example, the Mega-Float Phase 1 model has a length of 300 m, a width of 60 m and a depth of 2 m, while the Mega-Float Phase 2 model (Figure 3.1) has a length of 1,000 m, a width of 121 m (at the widest part) and a depth of 3 m. Given such a small depth to length ratio, the VLFS would behave almost like an elastic plate when considering the vertical responses. In other words, the VLFS response in the vertical direction cannot be assessed only by its rigid-body motions. It is therefore crucial to estimate the flexural response of a VLFS when designing the structure. However, for the design of the mooring system, the VLFS may be considered as a rigid-body since it has a large flexural rigidity in the horizontal directions.

When the dynamic response of a floating body due to waves is analyzed, the pressure of the fluid acting on the floating body must be evaluated at the same time. If a floating body moves, the fluid surrounding the floating body also moves. Thus, the pressure surrounding the floating body changes in order to satisfy the Bernoulli's equation. On the other hand, if the pressure changes, the motion of the floating body is affected. This kind of mutual relationship is referred to as a fluid-structure interaction. If the motion of the floating body consists of elastic deformations, the fluid-structure inter-action is called hydroelasticity. The vertical response of a VLFS typically shows hydroelasticity between its flexural behavior and the fluid motions. The hydroelastic analysis is thus necessary in order to assess the vertical response of a VLFS.

Many papers on hydroelastic analysis of VLFSs have been published to date. They may be found in the review papers by Kashiwagi (2000a), Watanabe et al. (2004a,b), Newman (2005), Ohmatsu (2005) and Suzuki et al. (2006). To name a few, Mamidipudi and Webster (1994) under-took pioneering work on the hydroelastic analysis of a mat-like floating

Figure 3.1 Mega-Float Phase 2 model moored in Tokyo Bay.
Source: Courtesy of Ship Research Center of Japan.

airport by combining the finite-difference method for plate problem and the Green's function method for fluid problem. Wu et al. (1995) solved the two dimensional (2-D) hydroelastic problem by the analytical method using eigenfunctions. Yago and Endo (1996) analyzed a zero-draft VLFS using the direct method and also compared this with their experimental results. Ohkusu and Nanba (1996) analyzed an infinite-length VLFS analytically. Kashiwagi (1998a) applied the B-spline panels for the analysis of a zero-draft VLFS using the pressure distribution method. Nagata et al. (1998) and Ohmatsu (1998) analyzed a rectangular VLFS by using a semi-analytical approach based on eigenfunction expansions in the depth direction. Seto and Ochi (1998) developed a hybrid element method (as a combination of the Finite Element Method (FEM) and infinite element) for hydroelastic analysis of a VLFS in stepped-depth configuration. Iijima et al. (1998) analyzed the hydroelastic behavior of semi-submersible type VLFSs using their program named VODAC. Takagi et al. (2000) analyzed the performance of an anti-motion device for a VLFS using the eigenfunction expansion method.

The transient response of a VLFS due to airplane landings and takeoffs was analyzed by Kim and Webster (1998), Watanabe and Utsunomiya (1996), and then for a moving load by Watanabe et al. (1998), Endo (2000) and Kashiwagi (2000b).

In the hydroelastic analysis of a VLFS, there are two major issues to be tackled. The first issue is concerned with the development of an accurate method for fluid-structure interaction analysis of an elastic floating vessel. In other words, how to model and analyze accurately the hydroelastic behavior of an elastic floating structure? The second issue deals with the applicability of the method to a VLFS. This is an important point of view because relatively large computational resources are required when analyzing a "very large" floating structure. Any established method for the hydroelastic analysis of VLFS must adequately satisfy these two requirements.

One way to tackle this problem is to use an analytical approach. If the problem has been solved analytically, the computational requirement for a VLFS is a non-issue. However, the drawback of the analytical approach is that only simple geometries are tractable. For example, modeled as a 2-D problem a floating plate, a circular plate, a ring-shaped plate or a rectangular plate can be solved analytically/semi-analytically. In these analytical/semi-analytical approaches, the constant sea depth is assumed. These analytical/semi-analytical approaches can be powerful tools, when the fundamental behavior of a VLFS is examined for scientific purposes in order to obtain a deeper knowledge or for preliminary design purposes. More importantly, the analytical/semi-analytical method can provide benchmark solutions for the verification of numerical solutions. In order to explain this analytical/semi-analytical approach in greater detail, the eigenfunction expansion matching method for a floating elastic plate modeled as a 2-D problem, which was first solved by Wu et al. (1995), is presented in the next section.

Another way to solve the hydroelastic problem of a VLFS is by using a numerical approach. In a numerical approach, there can be several combinations of the methods for solving the structure part and the fluid part. In principle, any method that can be applied to solve the vibration problem of a plate or a 3-D structure can be used for solving the structure part. Among these various methods, FEM may be considered as the most promising method because of its versatility in handling complicated geometries of real structures. For a linear hydroelastic analysis, commercial FEM codes may be used for modeling the VLFS's vessel structure. In some cases, the entire VLFS can be modeled as a floating plate. At the preliminary design stage, this simple modeling is very efficient and is frequently employed in the actual design procedure. For solving the fluid part, the often used method is the Green's function method. If one adopts Green's function which satisfies the boundary condition on the free-surface, the sea bottom and that at infinite distance from the floating structure, the unknown parameters to be determined for the fluid part can be minimized to be only those associated with the wetted surface of the floating body. At the same time, general geometry of the floating body and general topography of the seabed (including breakwater effect) can be considered. Alternatively, FEM can also be used for solving the fluid part although there are some difficulties encountered in satisfying accurately the infinite boundary condition and the whole fluid domain within the fictitious infinite boundary must be discretized into finite elements.

In order to couple the problem between the structure part and the fluid part, there exist two competing approaches. These two approaches are sometimes referred to as the modal method and the direct method. In the modal method, the flexural response of the structure is represented by a combination of the global modal responses of the structure. If each modal response corresponds to the natural mode for the vibration of the structure in air, the method may be called the dry-mode superposition method. On the other hand, if each modal response corresponds to the natural mode for vibration of the structure in fluid, the method may be referred to as the wet-mode superposition method. In actual applications, the dry-mode superposition method is frequently used since the dry-modes can be obtained in a straightforward manner by any standard FEM package. In some cases, the direct method is used. In the direct method, the response of the structure is not represented through a superposition of the global modal responses. Instead, the response of the structure is directly represented by, for example, the nodal responses if the structure is modeled by FEM. In most cases, the direct method is computationally more demanding than the modal method. Thus, nowadays the dry-mode superposition method seems more frequently used than the direct method for the actual design of a VLFS.

In Section 3.3, the modal method is explained in detail as an example of the numerical approach, where the classical thin-plate theory is employed for modeling the structural behavior and Green's function method is used for solving the fluid problem. This method can be used for a relatively large problem, if a computer equipped with large storage is available. However, a "very large" floating structure may be still difficult to solve because of its large requirements for storage and computational time. Fortunately, there exist several methods to overcome this difficulty. These methods will be explained briefly in the same section. Finally, some numerical results of a VLFS in a variable-depth sea are presented.

3.2   Hydroelastic analysis of a floating 2-D plate

3.2.1   Basic assumptions

The configuration and the adopted coordinate system for the analysis are shown in Figure 3.2. The 2-D sectional model assumes that the fluid and the structural motions are constant in the y-direction. A zero-draft plate floating

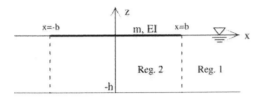

Figure 3.2 Configuration of the analytical model.

on a constant depth sea of h is assumed. The plate has a total length of $2b$, a mass per unit length m, and a constant bending rigidity EI.

By assuming irrotational motions of the perfect fluid, the boundary value problem for the total velocity potential $\Phi(x, z, t)$ is formulated. Considering steady state motions of the harmonically excited system at the circular frequency ω, the following expressions can be made:

$$\Phi(x, z, t) = \text{Re}\big[(\phi_D(x, z) + \phi_R(x, z))e^{-i\omega t}\big] \tag{3.1}$$

$$W(x, t) = \text{Re}\big[w(x)e^{-i\omega t}\big] \tag{3.2}$$

where t is the time, i the imaginary unit; $\phi_D(x, z)$ the diffraction potential, $\phi_R(x, z)$ the radiation potential; $W(x, t)$ the vertical deflection of the floating plate, and $w(x)$ the complex amplitude. The diffraction potential is obtained as the sum of the incident wave potential and the scattering wave potential for a floating body at rest. In this section, the incident wave is assumed to propagate from the positive x-direction along the x-axis to the negative x-direction. The radiation potential is the potential generated by the oscillating floating plate in still water (i.e. without any incident wave). Assuming the problem is linear, the total velocity potential can be represented by a linear superposition of the diffraction and radiation potentials. Note that once the complex velocity potentials $\phi_D(x, z)$ and $\phi_R(x, z)$ and the complex amplitude for the vertical deflection $w(x)$ are determined, the hydroelastic problem for this floating plate is completely solved.

3.2.2   Radiation potentials

The radiation potentials will be derived first. In order to account for the elastic deformation of the plate, the mode-expansion method is employed (see Newman 1994). As expressed by Eq. (3.3), the deflection of the plate $w(x)$ is represented by a superposition of arbitrarily chosen modal functions $f_l^S(x)\,(l = 1, \ldots, N_S)$ and $f_l^A(x)\,(l = 1, \ldots, N_A)$ including both rigid body motions and elastic deformations with modal amplitudes ζ_l^S and ζ_l^A, respectively. Here, the suffixes S and A indicate the symmetry and anti-symmetry of the solution with respect to the yz-plane, respectively. The total radiation potential $\phi_R(x, z)$ is then decomposed using the same modal amplitudes ζ_l^S and ζ_l^A and the corresponding unit-amplitude radiation potentials $\phi_l^S(x, z)$ and $\phi_l^A(x, z)$ are given by

$$w(x) = \sum_{l=0}^{N_S} \zeta_l^S f_l^S(x) + \sum_{l=0}^{N_A} \zeta_l^A f_l^A(x) \tag{3.3}$$

$$\phi_R(x, z) = -i\omega \left[\sum_{l=0}^{N_S} \zeta_l^S \phi_l^S(x, z) + \sum_{l=0}^{N_A} \zeta_l^A \phi_l^A(x, z) \right] \tag{3.4}$$

In the case of a beam model with free ends, Newman (1994) suggested that two sets of modal functions be used as the "generalized modes," that is, one set of natural modes of the beam and the other set comprising the Legendre polynomials. Because the natural modes (dry-modes) satisfy the free-end condition of the beam and provide a more effective expansion, they are used here (see Wu et al. 1995), that is

$$f_l^S(x) = \begin{cases} \dfrac{1}{2} & \text{for } l = 0 \\ \dfrac{1}{2} \left[\dfrac{\cosh(\mu_l^S x/b)}{\cosh \mu_l^S} + \dfrac{\cos(\mu_l^S x/b)}{\cos \mu_l^S} \right] & \text{for } l = 1, 2, \ldots, N_S \end{cases} \tag{3.5}$$

and

$$f_l^A(x) = \begin{cases} \dfrac{\sqrt{3}}{2} \dfrac{x}{b} & \text{for } l = 0 \\ \dfrac{1}{2} \left[\dfrac{\sinh(\mu_l^A x/b)}{\sinh \mu_l^A} + \dfrac{\sin(\mu_l^A x/b)}{\sin \mu_l^A} \right] & \text{for } l = 1, 2, \ldots, N_A \end{cases} \tag{3.6}$$

where $f_0^S(x)$ and $f_0^A(x)$ are the modes corresponding to the rigid-body motions, namely heave and pitch, respectively; $f_l^S(x)$ and $f_l^A(x)$ ($l = 1, 2, \ldots$) are the bending modal functions and μ_l^S and μ_l^A are the positive real roots of the equations:

$$\tan \mu_l^S + \tanh \mu_l^S = 0 \tag{3.7}$$

and

$$\tan \mu_l^A - \tanh \mu_l^A = 0 \tag{3.8}$$

The modal functions expressed in Eqs (3.5) and (3.6) are orthogonal to each other in the interval $(-b, b)$, that is,

$$\int_{-b}^{b} f_l^{S,A}(x) f_m^{S,A}(x)\, \mathrm{d}x = \frac{b}{2} \delta_{lm} \tag{3.9}$$

where δ_{lm} represents the Kronecker's delta function.

The boundary value problems for the unit-amplitude radiation potentials $\phi_l^S(x, z)$ and $\phi_l^A(x, z)$ are then formulated as follows:

$$\nabla^2 \phi_l^{S,A} = 0, \quad \text{in the fluid} \tag{3.10}$$

$$\frac{\partial \phi_l^{S,A}}{\partial z} = \frac{\omega^2}{g} \phi_l^{S,A}, \quad \text{on } z = 0,\ |x| > b \tag{3.11}$$

$$\frac{\partial \phi_l^{S,A}}{\partial z} = f_l^{S,A}(x), \quad \text{on } z = 0,\ |x| \le b \tag{3.12}$$

$$\frac{\partial \phi_l^{S,A}}{\partial z} = 0, \quad \text{on } z = -h \tag{3.13}$$

$$\phi_l^{S,A} \sim -i \frac{g A_l^{\pm}}{\omega} \frac{\cosh[k(z + h)]}{\cosh(kh)} e^{ik|x|}, \quad \text{as } x \to \pm\infty \tag{3.14}$$

where g is the gravitational acceleration, k the wave number, and $A_l^{\pm}$ the complex amplitudes of the radiated waves propagating away to infinity as $x \to \pm\infty$. The wave number k satisfies the following dispersion relationship:

$$k \tanh(kh) = \frac{\omega^2}{g} \tag{3.15}$$

Referring to McIver (1985), Wu et al. (1995) and Watanabe et al. (2003), the general solutions for the boundary value problems can be represented for regions 1 and 2 (see Figure 3.2) as follows:

$$\phi_l^{S,A(1)} = \sum_{n=0}^{\infty} h A_{ln}^{S,A} e^{-k_n(x-b)} \psi_n(z) \tag{3.16}$$

$$\phi_l^{S(2)} = h B_{l0}^{S} \chi_0(z) + \sum_{n=1}^{\infty} h B_{ln}^{S} \frac{\cosh(p_n x)}{\cosh(p_n b)} \chi_n(z) + \tilde{\phi}_l^{S(2)}(x, z) \tag{3.17}$$

$$\phi_l^{A(2)} = h B_{l0}^{A} \frac{x}{b} \chi_0(z) + \sum_{n=1}^{\infty} h B_{ln}^{A} \frac{\sinh(p_n x)}{\sinh(p_n b)} \chi_n(z) + \tilde{\phi}_l^{A(2)}(x, z) \tag{3.18}$$

The foregoing solutions are defined only for $x \geq 0$. The solutions for $x < 0$ can be easily obtained through symmetry or anti-symmetry of the solutions:

$$\phi_l^S(x,y) = \phi_l^S(-x,y) \qquad \text{for } x < 0 \tag{3.19}$$

$$\phi_l^A(x,y) = -\phi_l^A(-x,y) \qquad \text{for } x < 0 \tag{3.20}$$

The potentials $\tilde{\phi}_l^{S(2)}$ and $\tilde{\phi}_l^{A(2)}$ represent arbitrarily chosen particular solutions (inhomogeneous solutions) corresponding to the unit-amplitude motion of each modal function. The particular solutions $\tilde{\phi}_l^{S(2)}$ and $\tilde{\phi}_l^{A(2)}$ corresponding to the modal functions of Eqs (3.5) and (3.6) are given by

$$\tilde{\phi}_0^{S(2)} = \frac{1}{4h}(z^2 + 2hz - x^2) \tag{3.21}$$

$$\tilde{\phi}_0^{A(2)} = \frac{\sqrt{3}x}{12bh}[3h^2 - 3(h+z)^2 + x^2] \tag{3.22}$$

$$\tilde{\phi}_l^{S(2)} = \frac{b}{2\mu_l^S}\left\{ \frac{\cos(\mu_l^S x/b)}{\cos\mu_l^S}\frac{\cosh[\mu_l^S(z+h)/b]}{\sinh[\mu_l^S h/b]} \right.$$
$$\left. -\frac{\cosh(\mu_l^S x/b)}{\cosh\mu_l^S}\frac{\cos[\mu_l^S(z+h)/b]}{\sin[\mu_l^S h/b]} \right\} \tag{3.23}$$

$$\tilde{\phi}_l^{A(2)} = \frac{b}{2\mu_l^A}\left\{ \frac{\sin(\mu_l^A x/b)}{\sin\mu_l^A}\frac{\cosh[\mu_l^A(z+h)/b]}{\sinh[\mu_l^A h/b]} \right.$$
$$\left. -\frac{\sinh(\mu_l^A x/b)}{\sinh\mu_l^A}\frac{\cos[\mu_l^A(z+h)/b]}{\sin[\mu_l^A h/b]} \right\} \tag{3.24}$$

It can be readily checked that Eqs (3.21)–(3.24) satisfy Eqs (3.10), (3.12), and (3.13). Also, the following relationships are used.

$$\psi_n(z) = \frac{\cos[k_n(z+h)]}{N_n}, \quad n = 0, 1, \ldots \tag{3.25}$$

$$N_n^2 = \frac{1}{2}\left[1 + \frac{\sin(2k_n h)}{2k_n h}\right], \quad n = 0, 1, \ldots \tag{3.26}$$

$$k_0 = -ik; \quad -k_n\tan(k_n h) = \frac{\omega^2}{g}, \quad k_n > 0, \, n = 1, 2, \ldots \tag{3.27}$$

$$\chi_n(z) = \frac{\cos[p_n(z+h)]}{M_n}, \quad n = 0, 1, \ldots \tag{3.28}$$

$$M_0^2 = 1; \quad M_n^2 = \frac{1}{2}, \quad n = 1, 2, \ldots \tag{3.29}$$

$$p_n = \frac{n\pi}{h}, \quad n = 0, 1, \ldots \tag{3.30}$$

The following orthogonal relations are also satisfied among the z-dependence functions:

$$\frac{1}{h} \int_{-h}^{0} \psi_m(z) \psi_n(z) \, dz = \delta_{mn} \tag{3.31}$$

$$\frac{1}{h} \int_{-h}^{0} \chi_m(z) \chi_n(z) \, dz = \delta_{mn} \tag{3.32}$$

The continuity of the potential ϕ_l^S on the boundary between regions 1 and 2 (on $x = b$) can be represented as

$$\sum_{n=0}^{\infty} A_{ln}^S \psi_n(z) = B_{l0}^S \chi_0(z) + \sum_{n=1}^{\infty} B_{ln}^S \chi_n(z) + \frac{1}{h} \tilde{\phi}_l^{S(2)}(b, z) \tag{3.33}$$

By applying $\frac{1}{h} \int_{-h}^{0} \cdots \chi_m(z) \, dz$ to both sides of Eq. (3.33), one obtains

$$B_{lm}^S = \sum_{n=0}^{\infty} C_{mn} A_{ln}^S - \frac{1}{h^2} \int_{-h}^{0} \tilde{\phi}_l^{S(2)}(b, z) \chi_m(z) \, dz \tag{3.34}$$

$$C_{mn} = \frac{1}{h} \int_{-h}^{0} \chi_m(z) \psi_n(z) \, dz \tag{3.35}$$

The continuity of the horizontal velocity $\partial \phi_l^S / \partial x$ on the boundary between the regions 1 and 2 (on $x = b$) furnishes

$$-\sum_{n=0}^{\infty} k_n h A_{ln}^S \psi_n(z) = \sum_{n=1}^{\infty} \chi_n(z) p_n h B_{ln}^S \tanh(p_n b) + \frac{\partial \tilde{\phi}_l^{S(2)}(b, z)}{\partial x} \tag{3.36}$$

By applying $\frac{1}{h} \int_{-h}^{0} \cdots \psi_m(z) dz$ to both sizes of Eq. (3.36), one obtains

$$-k_m h A_{lm}^S = \sum_{n=1}^{\infty} C_{nm} p_n h B_{ln}^S \tanh(p_n b) + \frac{1}{h} \int_{-h}^{0} \frac{\partial \tilde{\phi}_l^{S(2)}(b, z)}{\partial x} \psi_m(z) \, dz \tag{3.37}$$

By substituting Eq. (3.34) into Eq. (3.37) and deleting the coefficient B_{ln}^S, one obtains the following equation having only A_{ln}^S as the unknowns:

$$k_m b A_{lm}^S + \sum_{n=0}^{\infty} \alpha_{mn}^S A_{ln}^S = \sum_{j=1}^{\infty} \frac{C_{jm} p_j b \tanh(p_j b)}{b^2} \int_{-b}^{0} \tilde{\phi}_l^{S(2)}(b,z) \chi_j(z) \, dz$$

$$- \frac{1}{b} \int_{-b}^{0} \frac{\partial \tilde{\phi}_l^{S(2)}(b,z)}{\partial x} \psi_m(z) \, dz \tag{3.38}$$

$$\alpha_{mn}^S = \sum_{j=1}^{\infty} C_{jm} C_{jn} p_j b \tanh(p_j b) \tag{3.39}$$

Since the particular solutions $\tilde{\phi}_l^{S(2)}(x,z)$ are given as Eqs (3.21) and (3.23), the unknown parameters A_{ln}^S can be obtained by solving Eq. (3.38) with appropriate truncation of the infinite sums. Then by substituting A_{ln}^S into Eq. (3.34), one obtains B_{ln}^S. In other words, the unit-amplitude radiation potential $\phi_l^S(x,z)$ has been solved.

By applying the similar procedure to $\phi_l^A(x,z)$, one finally arrives at the following equation:

$$k_m b A_{lm}^A + \sum_{n=0}^{\infty} \alpha_{mn}^A A_{ln}^A = \frac{C_{0m}}{ab} \int_{-b}^{0} \tilde{\phi}_l^{A(2)}(b,z) \chi_0(z) \, dz$$

$$+ \sum_{j=1}^{\infty} \frac{C_{jm} p_j b \coth(p_j b)}{b^2} \int_{-b}^{0} \tilde{\phi}_l^{A(2)}(b,z) \chi_j(z) \, dz$$

$$- \frac{1}{b} \int_{-b}^{0} \frac{\partial \tilde{\phi}_l^{A(2)}(b,z)}{\partial x} \psi_m(z) \, dz \tag{3.40}$$

$$\alpha_{mn}^A = C_{0m} C_{0n} \frac{b}{b} + \sum_{j=1}^{\infty} C_{jm} C_{jn} p_j b \coth(p_j b) \tag{3.41}$$

3.2.3 Diffraction potentials

The diffraction potential can be obtained by the similar procedure used in the radiation potentials. The boundary value problem for the diffraction potential $\phi_D(x,z)$ may be formulated as follows (McIver 1985):

$$\nabla^2 \phi_D = 0, \quad \text{in the fluid} \tag{3.42}$$

$$\frac{\partial \phi_D}{\partial z} = \frac{\omega^2}{g} \phi_D, \quad \text{on } z = 0, \ |x| > b \tag{3.43}$$

$$\frac{\partial \phi_D}{\partial z} = 0, \quad \text{on } z = 0, \ |x| \leq b \tag{3.44}$$

$$\frac{\partial \phi_D}{\partial z} = 0, \quad \text{on } z = -h \tag{3.45}$$

$$\phi_D \sim \begin{cases} \phi_0 \cosh[k(z+h)](e^{-ikx} + Re^{ikx}) & \text{as } x \to \infty \\ \phi_0 T \cosh[k(z+h)]e^{-ikx} & \text{as } x \to -\infty \end{cases} \tag{3.46}$$

where ϕ_0 corresponding to the incident wave of the amplitude A is defined as

$$\phi_0 = -i\frac{gA}{\omega}\frac{1}{\cosh(kh)} \tag{3.47}$$

The diffraction potential may be expressed as the sum of the symmetric and the anti-symmetric parts:

$$\phi_D(x,z) = \phi_D^S(x,z) + \phi_D^A(x,z) \tag{3.48}$$

where

$$\phi_D^S(x,y) = \phi_D^S(-x,y), \quad \text{for } x < 0 \tag{3.49}$$

$$\phi_D^A(x,y) = -\phi_D^A(-x,y), \quad \text{for } x < 0 \tag{3.50}$$

Then, the general solutions for $\phi_D^S(x,z)$ and $\phi_D^A(x,z)$ can be represented for regions 1 and 2 (see Figure 3.2) as follows:

$$\phi_D^{S,A(1)} = \phi_0 N_0 \left[\frac{1}{2}e^{-ikx}\psi_0(z) + \sum_{n=0}^{\infty} A_{Dn}^{S,A} e^{-k_n(x-b)}\psi_n(z) \right] \tag{3.51}$$

$$\phi_D^{S(2)} = \phi_0 N_0 \left[B_{D0}^S \chi_0(z) + \sum_{n=1}^{\infty} B_{Dn}^S \frac{\cosh(p_n x)}{\cosh(p_n b)}\chi_n(z) \right] \tag{3.52}$$

$$\phi_D^{A(2)} = \phi_0 N_0 \left[B_{D0}^A \frac{x}{b}\chi_0(z) + \sum_{n=1}^{\infty} B_{Dn}^A \frac{\sinh(p_n x)}{\sinh(p_n b)}\chi_n(z) \right] \tag{3.53}$$

The radiation condition at infinity, as given by Eq. (3.46), is satisfied if

$$R = (A_{D0}^S + A_{D0}^A)e^{-ikb} \tag{3.54}$$

$$T = (A_{D0}^S - A_{D0}^A)e^{-ikb} \tag{3.55}$$

where $|R|$ and $|T|$ correspond to the reflection and transmission coefficients, respectively.

The continuity of the potential ϕ_D^S on the boundary between the regions 1 and 2 (on $x = b$) can be represented as

$$\frac{1}{2}e^{-ika}\psi_0(z) + \sum_{n=0}^{\infty} A_{Dn}^S \psi_n(z) = B_{D0}^S \chi_0(z) + \sum_{n=1}^{\infty} B_{Dn}^S \chi_n(z) \tag{3.56}$$

By applying $\frac{1}{b}\int_{-b}^{0} \cdots \chi_m(z)\,dz$ to both sides of Eq. (3.56), one obtains

$$B_{Dm}^S = \sum_{n=0}^{\infty} C_{mn} A_{Dn}^S + \frac{1}{2}e^{-ikb}C_{m0} \tag{3.57}$$

Then, the continuity of the horizontal velocity $\partial \phi_D^S / \partial x$ on the boundary between the regions 1 and 2 (on $x = b$) gives

$$-\sum_{n=0}^{\infty} k_n A_{Dn}^S \psi_n(z) = \sum_{n=1}^{\infty} \chi_n(z) p_n B_{Dn}^S \tanh(p_n b) \tag{3.58}$$

By applying $\int_{-b}^{0} \cdots \psi_m(z)\,dz$ to both sides of Eq. (3.58), one obtains

$$-k_m b A_{Dm}^S = \sum_{n=1}^{\infty} C_{nm} p_n b B_{Dn}^S \tanh(p_n b) \tag{3.59}$$

By substituting Eq. (3.57) into Eq. (3.59) and deleting the coefficient B_{Dn}^S, we obtain the following equation having only A_{Dn}^S as the unknowns:

$$k_m b A_{Dm}^S + \sum_{n=0}^{\infty} \alpha_{mn}^S A_{Dn}^S = -\frac{1}{2}e^{-ikb}\sum_{j=1}^{\infty} C_{jm}C_{j0}p_j b \tanh(p_j b) \tag{3.60}$$

In Eq. (3.60), the coefficient α_{mn}^S is given as Eq. (3.39). By solving Eq. (3.60), we obtain the unknown coefficient A_{Dn}^S; and then from Eq. (3.57), we obtain the coefficient B_{Dn}^S. Similarly, we can formulate similar equations for A_{Dn}^A and B_{Dn}^A as

$$B_{Dm}^A = \sum_{n=0}^{\infty} C_{mn} A_{Dn}^A + \frac{1}{2}e^{-ikb}C_{m0} \tag{3.61}$$

$$k_m b A_{Dm}^A + \sum_{n=0}^{\infty} \alpha_{mn}^A A_{Dn}^A = -\frac{1}{2}e^{-ikb}\sum_{j=1}^{\infty} C_{jm}C_{j0}p_j b \coth(p_j b) \tag{3.62}$$

By solving Eqs (3.61) and (3.62), one obtains the coefficients A_{Dn}^A and B_{Dn}^A, and finally the diffraction potential itself can be obtained as an analytical expression by using Eqs (3.48)–(3.53).

3.2.4 Equations of motion

The equations of motion in the generalized modal coordinates ζ_l are derived using the energy approach as shown below. The kinetic energy of the floating plate T is given by

$$T = \frac{1}{2}m\omega^2 \int_{-b}^{b} w(x)^2 \, dx \tag{3.63}$$

The sum of the strain energy U by the elastic deformation of the floating body and the energy induced by the hydrostatic restoring force is given by

$$U = \frac{1}{2}EI \int_{-b}^{b} w''(x)^2 dx + \frac{1}{2}\rho g \int_{-b}^{b} w(x)^2 dx \tag{3.64}$$

The potential energy induced by the external force which is independent of $w(x)$ is given by

$$V = -\int_{-b}^{b} p(x)w(x)dx \tag{3.65}$$

where $p(x)$ is the complex amplitude of the dynamic pressure $P(x,t)$ that acts on the bottom surface of the floating body. From the linearized Bernoulli's equation, the pressure on the bottom surface of the floating body $P(x,t)$ is related to the potential by

$$P(x,t) = \mathrm{Re}\left[p(x)e^{-i\omega t}\right] = -\rho \frac{\partial \Phi(x,0,t)}{\partial t} \tag{3.66}$$

Thus, the following relationship can be derived:

$$p(x) = \rho i\omega \phi_D^S(x,0) + \rho i\omega \phi_D^A(x,0) + \rho\omega^2 \sum_{j=0}^{N_S} \zeta_j^S \phi_j^S(x,0)$$

$$+ \rho\omega^2 \sum_{j=0}^{N_A} \zeta_j^A \phi_j^A(x,0) \tag{3.67}$$

By applying the Hamilton's principle, one obtains

$$-m\omega^2 \int_{-b}^{b} w(x)\delta w(x) \, dx + EI \int_{-b}^{b} w''(x)\delta w''(x) \, dx$$

$$+ \rho g \int_{-b}^{b} w(x)\delta w(x) \, dx - \int_{-b}^{b} p(x)\delta w(x) \, dx = 0 \tag{3.68}$$

We then substitute Eq. (3.3) for $w(x)$, Eq. (3.67) for $p(x)$, and Eq. (3.69) for $\delta w(x)$ in Eq. (3.68).

$$\delta w(x) = \sum_{l=0}^{N_S} \delta \zeta_l^S f_l^S(x) + \sum_{l=0}^{N_A} \delta \zeta_l^A f_l^A(x) \tag{3.69}$$

Taking into account the arbitrariness of $\delta \zeta_i$, one finally obtains the following equations of motion:

$$\sum_{j=0}^{N_S} \left[-\omega^2 \left\{ M_{ij}^S + M_{aij}^S \right\} + K_{ij}^S \right] \zeta_j^S = F_i^S, \quad i = 0, 1, \ldots, N_S \tag{3.70}$$

$$\sum_{j=0}^{N_A} \left[-\omega^2 \left\{ M_{ij}^A + M_{aij}^A \right\} + K_{ij}^A \right] \zeta_j^A = F_i^A, \quad i = 0, 1, \ldots, N_A \tag{3.71}$$

where the generalized mass $M_{ij}^{S,A}$, the generalized added mass $M_{aij}^{S,A}$, the generalized stiffness $K_{ij}^{S,A}$, and the generalized exciting force $F_i^{S,A}$ are given by:

$$M_{ij}^{S,A} = m \int_{-b}^{b} f_i^{S,A}(x) f_j^{S,A}(x) \, dx \tag{3.72}$$

$$K_{ij}^{S,A} = EI \int_{-b}^{b} \left(f_i^{S,A}(x) \right)'' \left(f_j^{S,A}(x) \right)'' \, dx + \rho g \int_{-b}^{b} f_i^{S,A}(x) f_j^{S,A}(x) \, dx \tag{3.73}$$

$$M_{aij}^{S,A} = \rho \int_{-b}^{b} \phi_j^{S,A}(x, 0) f_i^{S,A}(x) \, dx \tag{3.74}$$

$$F_i^{S,A} = \rho i \omega \int_{-b}^{b} \phi_D^{S,A}(x, 0) f_i^{S,A}(x) \, dx \tag{3.75}$$

The equations of motion given by Eqs (3.70) and (3.71), are uncoupled. Thus, they can be solved separately. By solving the equations of motion in the frequency domain for ζ_j^S and ζ_j^A, and using Eq. (3.3), the response of the elastic floating plate is obtained.

3.2.5 Analytical results

The analyzed model corresponds to the experimental model reported in Ohta et al. (1998). The specifications are $EI/\rho g b^2 = 2.93 \times 10^{-4}$, $m/\rho h = 0.0867$, the water depth $h = 1.1$ m, and the half length of the floating body $b = 5.0$ m. As an example of the analytical results, Figure 3.3 shows the vertical deflection of the floating body when $L/\lambda = 10.0$ ($L := 2b$ the total

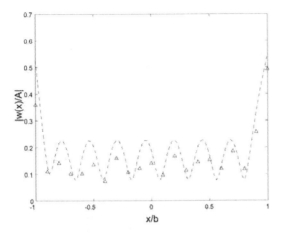

Figure 3.3 Response amplitude operator along the floating body.

Source: Watanabe et al. (2003).

Notes
Broken line: analytical results, triangles: experimental results.

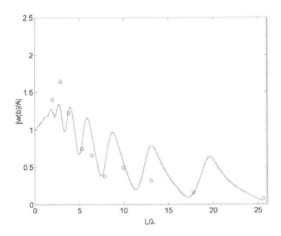

Figure 3.4 Response amplitude operator at fore-end ($x = b$).

Source: Watanabe et al. (2003).

Notes
Solid line: analytical results, circles: experimental results.

length of the floating body and λ is the incident wave length) together with the experimental results. The incident wave propagates along the negative x-direction. Basically, a good agreement is observed between the theoretical results and the experimental results.

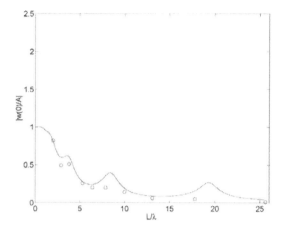

Figure 3.5 Response amplitude operator at mid-position ($x = 0$).

Source: Watanabe et al. (2003).

Notes
Solid line: analytical results, circles: experimental results.

Figures 3.4 and 3.5 show the vertical response at the fore-end and at the mid-position of the floating body as a function of L/λ. Satisfactory agreement of the analytical results with the experimental results was observed.

3.3 Hydroelastic analysis of a VLFS in variable water depth

3.3.1 Governing equations and boundary conditions

Consider a box-like VLFS of length $2a$ ($=L$), width $2b$ ($=B$), draft d, and floating in the open sea of variable depth (with a constant depth h at infinity) as shown in Figure 3.6. The variable depth sea-bottom surface (S_B) is defined such that the boundary line (Γ_B) touches on the flat bottom base-surface of $z = -h$ and S_B must be located higher than the base surface ($z = -h$). The coordinate system is defined such that the xy-plane lies on the undisturbed free surface and the z-axis points upwards. The center of the VLFS is on the z-axis as shown in Figure 3.6.

Consider long-crested harmonic waves with a small amplitude. The amplitude of the incident wave is defined by A, the circular frequency by ω, and the angle of incidence by β ($\beta = 0$ corresponds to the head wave from positive x-direction and $\beta = \pi/2$ to the beam wave from positive y-direction).

By assuming the water to be a perfect fluid with no viscosity and incompressible, and the fluid motion to be irrotational, the fluid motion can be

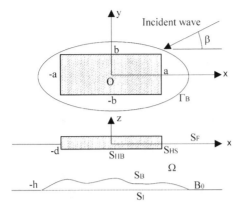

Figure 3.6 Configuration of the analytical model.

represented by a velocity potential Φ. Also, we consider the steady-state harmonic motions of the fluid and the structure, with the circular frequency ω. Then, all of the time-dependent quantities can be represented similarly as follows:

$$\Phi(x, y, z; t) = \text{Re}\left[\phi(x, y, z)e^{i\omega t}\right] \tag{3.76}$$

where i is the imaginary unit and t the time. In the following, all time-dependent variables are represented in the frequency domain unless stated explicitly.

It is also assumed that the fluid and structural motions are small so that the linear potential theory can be applied for formulating the fluid-motion problem. Based on these assumptions, the following boundary value problem may be formulated:

$$\nabla^2\phi = 0 \quad \text{in } \Omega \tag{3.77}$$

$$\frac{\partial\phi}{\partial z} = K\phi \quad \text{on } S_F \tag{3.78}$$

$$\frac{\partial\phi}{\partial n} = 0 \quad \text{on } S_B \tag{3.79}$$

$$\frac{\partial\phi}{\partial z} = 0 \quad \text{on } B_0 \tag{3.80}$$

$$\frac{\partial\phi}{\partial n} = i\omega w(x, y) \quad \text{on } S_{HB} \tag{3.81}$$

$$\frac{\partial\phi}{\partial n} = 0 \quad \text{on } S_{HS} \tag{3.82}$$

$$\lim_{r \to \infty} \sqrt{r} \left(\frac{\partial(\phi - \phi_I)}{\partial r} - ik(\phi - \phi_I) \right) = 0 \quad \text{on } S_\infty \tag{3.83}$$

$$\phi_I = i \frac{gA}{\omega} \frac{\cosh k(z + h)}{\cosh kh} e^{ik(x \cos \beta + y \sin \beta)} \tag{3.84}$$

where ϕ_I is the incident wave potential. The symbols Ω, S_F, S_B, and B_0 represent the fluid domain, the free surface, the variable depth sea-bottom surface, and the flat-bottom base surface on $z = -h$, respectively. The symbols S_{HB}, S_{HS}, and S_∞ represent the bottom surface of the floating body, the wetted side surface of the floating body, and the artificial fluid boundary at infinity, respectively. The symbol K represents the wave number at infinite depth sea ($=\omega^2/g$ and g is the gravitational acceleration), whereas k is the wave number at constant depth h satisfying the following dispersion relation

$$k \tanh kh = K \tag{3.85}$$

The symbol n represents a unit normal vector (the positive direction points out of the fluid domain). The complex-valued variable $w(x, y)$ represents the vertical deflection of the bottom surface of the floating body. The variable r in Eq. (3.83) denotes the horizontal distance between the origin and the referred point. Note that the shallow draft assumption of the floating body is not employed in the formulation although the flat-bottom of the floating body is assumed in this formulation.

It is now widely accepted that VLFS response in terms of the vertical deflection can be captured well by modeling the whole VLFS as an elastic plate. In this formulation, the classical thin plate theory on elastic foundations (which model the hydrostatic restoring forces) is adopted:

$$D\nabla^4 w(x, y) - \omega^2 \gamma w(x, y) + kw(x, y) = p(x, y) \tag{3.86}$$

where D is the plate rigidity, γ is the mass per unit area of the plate, $k = \rho g$ (ρ: density of fluid), and $p(x, y)$ is the dynamic pressure on the bottom surface of the VLFS. The dynamic pressure $p(x, y)$ relates to the velocity potential on the bottom surface of the VLFS from the linearized Bernoulli's equation

$$p(x, y) = -i\rho\omega\phi(x, y, -d) \tag{3.87}$$

The floating body is not constrained in the vertical direction and along its edges, the following boundary conditions for a free edge must be satisfied:

$$\left[\frac{\partial^3 w(x, y)}{\partial x^3} + (2 - v) \frac{\partial^3 w(x, y)}{\partial x \partial y^2} \right]_{x=\pm a} = 0 \tag{3.88}$$

$$\left[\frac{\partial^2 w(x, y)}{\partial x^2} + v \frac{\partial^2 w(x, y)}{\partial y^2} \right]_{x=\pm a} = 0 \tag{3.89}$$

where v is Poisson's ratio. The boundary conditions along the edges $y = \pm b$ are given similarly.

3.3.2 Modal expansion of the motion and the potential

The plate equation (3.86) indicates that the response of the plate, $w(x, y)$, is coupled with the fluid motions (or velocity potential, $\phi(x, y, z)$) through Eq. (3.87). On the other hand, the fluid motion can be obtained only when the plate response $w(x, y)$ is specified in the boundary condition, Eq. (3.81). In order to decouple this interaction problem into the hydrodynamic problem in terms of the velocity potential and the mechanical problem for the vibration of the plate, the motion of the plate $w(x, y)$ is expanded by a series of the products of the modal functions $f_l(x, y)$ and the complex amplitudes ζ_l:

$$w(x, y) = \sum_{l=1}^{P} \zeta_l f_l(x, y) \tag{3.90}$$

The potential $\phi(x, y, z)$ can be expressed by the sum of the diffraction and radiation potentials, $\phi_D(x, y, z)$ and $\phi_R(x, y, z)$, based on the linear theory, and the radiation potential $\phi_R(x, y, z)$ can be further decomposed as follows (Eatock Taylor & Waite 1978; Newman 1994):

$$\phi(x, y, z) = \phi_D(x, y, z) + i\omega \sum_{l=1}^{P} \zeta_l \phi_l(x, y, z) \tag{3.91}$$

where $\phi_l(x, y, z)$ is the radiation potential corresponding to the unit-amplitude motion of the l-th modal function. The diffraction potential $\phi_D(x, y, z)$ may be further decomposed as

$$\phi_D(x, y, z) = \phi_I(x, y, z) + \phi_S(x, y, z) \tag{3.92}$$

where $\phi_S(x, y, z)$ is the scattering potential.

By substituting Eqs (3.90) and (3.91) into Eqs (3.77)–(3.83), we have the following governing equation and boundary conditions for each of the unit-amplitude radiation potentials (for $l = 1, 2, \ldots, P$) and the diffraction potential (for $l = D$).

$$\nabla^2 \phi_l = 0 \quad \text{in } \Omega \tag{3.93}$$

$$\frac{\partial \phi_l}{\partial z} = K\phi_l \quad \text{on } S_F \tag{3.94}$$

$$\frac{\partial \phi_l}{\partial n} = 0 \quad \text{on } S_B \tag{3.95}$$

$$\frac{\partial \phi_l}{\partial z} = 0 \quad \text{on } B_0 \tag{3.96}$$

$$\frac{\partial \phi_l}{\partial n} = \begin{cases} f_l(x, y) & \text{for } l = 1, 2, \ldots, P \\ 0 & \text{for } l = D \end{cases} \quad \text{on } S_{HB} \tag{3.97}$$

$$\frac{\partial \phi_l}{\partial n} = 0 \quad \text{on } S_{HS} \tag{3.98}$$

$$\lim_{r \to \infty} \sqrt{r} \left(\frac{\partial \phi_l}{\partial r} - ik\phi_l \right) = 0 \quad \text{for } l = 1, 2, \ldots, P, S \text{ on } S_{\infty} \tag{3.99}$$

Now, the boundary value problems for each of the unit-amplitude radiation potentials and the diffraction potential are given explicitly in an uncoupled form.

3.3.3   Hydrodynamic analysis

The boundary value problems given by Eqs (3.93)–(3.99) can be solved by using the Green's function method. If the free-surface Green's function satisfying the boundary conditions given by Eqs (3.94), (3.96) and (3.99) is used, the boundary integral equation for each of the unknown potential can be derived as follows:

$$C(\vec{x})\phi_l(\vec{x}) + \int_{S_{HB} \cup S_{HS} \cup S_B} \frac{\partial G(\vec{x}, \vec{\xi})}{\partial n} \phi_l(\vec{\xi}) \, dS_{\xi} = \int_{S_{HB} \cup S_{HS} \cup S_B} G(\vec{x}, \vec{\xi}) \frac{\partial \phi_l(\vec{\xi})}{\partial n} \, dS_{\xi} \tag{3.100}$$

Further, by introducing the boundary conditions for $\partial \phi_l / \partial n$ into Eq. (3.100), we have

$$C(\vec{x})\phi_l(\vec{x}) + \int_{S_{HB} \cup S_{HS} \cup S_B} \frac{\partial G(\vec{x}, \vec{\xi})}{\partial n} \phi_l(\vec{\xi}) \, dS_{\xi}$$

$$= \begin{cases} \int_{S_{HB}} G(\vec{x}, \vec{\xi}) f_l(\vec{\xi}) \, dS_{\xi} & \text{for } l = 1, 2, \ldots, P \\ 4\pi \phi_0(\vec{x}) & \text{for } l = D \end{cases} \tag{3.101}$$

Here, $\vec{x} = (x, y, z)$ and $\vec{\xi} = (\xi, \eta, \zeta)$. The function $G(\vec{x}, \vec{\xi})$ denotes the Green's function representing water waves, and it satisfies the boundary conditions on the free surface, on the flat sea bottom $(z = -h)$, and the radiation condition at infinity. The series form can be represented as follows (John 1950; Newman 1985):

$$G(\vec{x},\vec{\xi}) = \sum_{m=0}^{\infty} \frac{2K_0(k_m R)}{N_m} \cos k_m(z+h) \cos k_m(\zeta+h) \tag{3.102}$$

$$N_m = \frac{h}{2}\left(1 + \frac{\sin 2k_m h}{2k_m h}\right) \tag{3.103}$$

$$k_m \tan k_m h = -K \tag{3.104}$$

where k_m ($m \geq 1$) is a positive root number, and $k_0 = ik$. The following relationship is also used

$$K_0(ikR) = -\tfrac{1}{2}\pi i H_0^{(2)}(kR) \tag{3.105}$$

where K_0 is the modified Bessel function of the second kind, and $H_0^{(2)}$ is the Hankel function. The symbol R represents horizontal distance between $\vec{x}$ and $\vec{\xi}$.

The integral equation (3.101) may be solved by using the conventional constant panel method, in which case the coefficient C becomes constant and the singular integrals may be solved analytically. However, if a higher order BEM is applied to solve Eq. (3.101), the evaluation of C, which corresponds to the solid angle over which the fluid is viewed from $\vec{x}_0$, and the numerical integration of the singular integrals appear to be difficult to perform. In order to overcome this difficulty, alternative integral equation (Teng and Eatock Taylor 1995) may be used. This is of the form:

$$\left(4\pi + \int_{S_I} \frac{\partial G_2(\vec{x},\vec{\xi})}{\partial z}\, dS_\xi\right) \phi_l(\vec{x})$$

$$+ \int_{S_{HB} \cup S_{HS} \cup S_B} \left\{\phi_l(\vec{\xi})\frac{\partial G(\vec{x},\vec{\xi})}{\partial n} - \phi_l(\vec{x})\frac{\partial G_2(\vec{x},\vec{\xi})}{\partial n}\right\} dS_\xi$$

$$= \begin{cases} \int_{S_{HB}} G(\vec{x},\vec{\xi})f_l(\vec{\xi})dS_\xi & \text{for } l = 1, 2, \ldots, P \\ 4\pi\phi_0(\vec{x}) & \text{for } l = D \end{cases} \tag{3.106}$$

$$G_2(\vec{x},\vec{\xi}) = \frac{1}{r} + \frac{1}{r_1} + \frac{1}{r_{21}} + \frac{1}{r_{31}} \tag{3.107}$$

$$r = [R^2 + (z-\zeta)^2]^{1/2}$$

$$r_1 = [R^2 + (z+\zeta)^2]^{1/2}$$

$$r_{21} = [R^2 + (z-\zeta-2h)^2]^{1/2}$$

$$r_{31} = [R^2 + (z+\zeta+2h)^2]^{1/2}$$

The symbol S_I designates the inner plane of $z = -h$ inside the boundary Γ_B. Using Eq. (3.106), the evaluations of solid angles C and CPV integrals can be avoided. The integral equation given by Eq. (3.106) can be solved for each of velocity potentials ϕ_l by discretizing the surface of the boundaries S_{HB}, S_{HS}, S_B and S_I into panels using a standard procedure known as the boundary element method (BEM).

3.3.4   Equations of motion

The Hamilton's principle is used to derive the equations of motion:

$$\delta(U - T + V) = 0 \tag{3.108}$$

where U represents the strain energy, T the kinetic energy, and V the potential energy due to external force. These energies are given by

$$U = \frac{1}{2} \int_{S_{HB}} \left[D \left\{ \left(\frac{\partial^2 w}{\partial x^2} \right)^2 + \left(\frac{\partial^2 w}{\partial y^2} \right)^2 + 2\nu \frac{\partial^2 w}{\partial x^2} \frac{\partial^2 w}{\partial y^2} \right. \right.$$

$$\left. \left. +2(1 - \nu) \left(\frac{\partial^2 w}{\partial x \partial y} \right)^2 \right\} + kw^2 \right] \mathrm{d}S \tag{3.109}$$

$$T = \frac{1}{2} \omega^2 \gamma \int_{S_{HB}} w^2 \, \mathrm{d}S \tag{3.110}$$

$$V = - \int_{S_{HB}} p(x, y) w \, \mathrm{d}S \tag{3.111}$$

By substituting Eqs (3.87), (3.91) and (3.105)–(3.111) into Eq. (3.108), and taking variations with respect to w, we have

$$D \int_{S_{HB}} \left[\frac{\partial^2 w}{\partial x^2} \frac{\partial^2 \delta w}{\partial x^2} + \frac{\partial^2 w}{\partial y^2} \frac{\partial^2 \delta w}{\partial y^2} + \nu \frac{\partial^2 w}{\partial x^2} \frac{\partial^2 \delta w}{\partial y^2} + \nu \frac{\partial^2 w}{\partial y^2} \frac{\partial^2 \delta w}{\partial x^2} \right.$$

$$\left. +2(1 - \nu) \frac{\partial^2 w}{\partial x \partial y} \frac{\partial^2 \delta w}{\partial x \partial y} \right] \mathrm{d}S + k \int_{S_{HB}} w \delta w \, \mathrm{d}S - \omega^2 \gamma \int_{S_{HB}} w \delta w \, \mathrm{d}S$$

$$= -i\rho\omega \int_{S_{HB}} \phi_D(x, y, -d) \delta w \, \mathrm{d}S + \rho\omega^2 \sum_{l=1}^{P} \varsigma_l \int_{S_{HB}} \phi_l(x, y, -d) \delta w \, \mathrm{d}S \tag{3.112}$$

Here, substituting Eq. (3.90) and the variation into Eq. (3.112), and considering the arbitrariness of $\delta\zeta_l$, one finally obtains the following equation of motion in the modal coordinates:

$$\sum_{l=1}^{P} \zeta_l \{K_{lj} - \omega^2 (M_{lj} + M_{a_{lj}})\} = F_j \tag{3.113}$$

$$K_{lj} = D \int_{S_{HB}} \left[\frac{\partial^2 f_l}{\partial x^2} \frac{\partial^2 f_j}{\partial x^2} + \frac{\partial^2 f_l}{\partial y^2} \frac{\partial^2 f_j}{\partial y^2} + v\frac{\partial^2 f_l}{\partial x^2} \frac{\partial^2 f_j}{\partial y^2} \right.$$
$$\left. + v\frac{\partial^2 f_l}{\partial y^2} \frac{\partial^2 f_j}{\partial x^2} + 2(1-v)\frac{\partial^2 f_l}{\partial x \partial y} \frac{\partial^2 f_j}{\partial x \partial y} \right] dS + k \int_{S_{HB}} f_l f_j \, dS \tag{3.114}$$

$$M_{lj} = \gamma \int_{S_{HB}} f_l f_j \, dS \tag{3.115}$$

$$M_{alj} = \rho \int_{S_{HB}} \phi_l(x,y,-d) f_j \, dS \tag{3.116}$$

$$F_j = -i\rho\omega \int_{S_{HB}} \phi_D(x,y,-d) f_j \, dS \tag{3.117}$$

where K_{lj} is the generalized stiffness matrix, M_{lj} the generalized mass matrix, M_{alj} the generalized added-mass matrix, and F_j the external force vector. The generalized added-mass matrix is given as a complex matrix, thus including the effect of the radiation damping. The selection of the modal function f_j has in fact some arbitrariness; for examples, a tensor product of modal functions of a vibrating free-free beam in the x- and y-directions may be used (Utsunomiya et al. 1998). When the structure is modeled by FEM, the dry-modes of the plate would be most convenient. In this case, the generalized equation of motion (3.113) can be simplified to

$$\sum_{l=1}^{P} \zeta_l \{\omega_j^2 - \omega^2 (\delta_{lj} + M_{a_{lj}})\} = F_j \tag{3.118}$$

where ω_j is the j-th eigen-frequency and δ_{lj} the Kronecker's delta function. By solving either Eqs (3.113) or (3.118), the modal amplitudes ζ_l are obtained. By substituting these modal amplitudes ζ_l into Eq. (3.90), one finally obtains the deflection of the VLFS. The pressure acting on the VLFS can be determined by Eqs (3.87) and (3.91) where the modal amplitudes ζ_l are substituted. These values will be used to check the safety and serviceability of the VLFS. The stress-resultants, *viz.* the bending moments, the torsional moments and the shearing forces, can be obtained through the modal analysis, where the stress-resultants are calculated in advance for each dry-mode, and the final values are obtained by the modal superposition. These values will be used for stress analysis using a 3-D detailed FEM model of a part of VLFS (Inoue et al. 2003; Fujikubo 2005).

Although the above-mentioned approach is given for modeling the whole VLFS as an isotropic plate, the extension to an orthotropic plate or a 3-D general model is straightforward.

3.3.5   Comments on direct method

In the direct method, the equations of motion also appear in the same form as Eq. (3.113). However, the modal response ζ_l would be replaced by the nodal response (i.e. the response at the nodal points of the structure) w_l. Thus, the stiffness and mass matrices come directly from the FEM model of the structure. Naturally, the added-mass matrix should have the same degree of freedoms as the number of nodes (or total degree of freedom of the nodes) located on the wetted-surface of the floating body. Although this would increase the computational time and storage requirement, the analyst need not worry about the convergence on the number of modal functions to be used. For example, Yago and Endo (1996) and Kashiwagi (1998b) used the direct method to solve the hydroelastic response of VLFS in regular waves.

3.3.6   Accelerated methods for very large hydroelastic problems

As mentioned earlier, the hydroelastic behavior of a VLFS may be analyzed by using Green's function method for solving the fluid part. However, the Green's function method produces a complex-valued full matrix of $O(N^2)$ ($N :=$ number of nodes in the BEM). In the following, the required size of the matrix for a typical VLFS problem is treated.

A wetted-body surface must be discretized into meshes, where the size of the meshes should be small enough when compared to the incident wave length. Experience shows that for a constant panel, the mesh size should be less than 1/10 of the incident wave length. For an 8-node quadratic panel, it should be less than 1/4 of the incident wave length while for a B-spline panel, the required number of unknowns may be estimated from the work by Kashiwagi (1998a). As typical examples, we consider Model A (a commuter airport class: $L \times B \times d = 2{,}000\,\text{m} \times 400\,\text{m} \times 2\,\text{m}$; L: length, B: width, d: draft), Model B (an international airport class: $L \times B \times d = 5{,}000\,\text{m} \times 1{,}000\,\text{m} \times 4\,\text{m}$) and Model C (model B with variable water depth for the seabed area of $6{,}000\,\text{m} \times 2{,}000\,\text{m}$). When the VLFS models are analyzed for the incident wave length of $\lambda = 50\,\text{m}$, the required panel sizes are $5\,\text{m} \times 5\,\text{m}$ for a constant panel, and $12.5\,\text{m} \times 12.5\,\text{m}$ for an 8-node quadratic panel. The required number of unknowns and the storage size for the system matrix (estimated by (number on unknowns)$^2 \times 16$ bytes) are summarized in Table 3.1.

Table 3.1 Storage requirement in Green's function method

Model	Constant panel		8-node quadratic panel			B-spline panel	
	Number of panels	Storage	Number of panels	Number of unknowns	Storage	Number of unknowns	Storage
A	32,960	16.2 GB	5,504	16,897	4.25 GB	3,125	149 MB
B	212,000	670 GB	32,960	99,841	149 GB	15,245	3.46 GB
C	692,000	6.97 TB	109,760	331,522	1.60 TB	46,454	32.2 GB

Obviously, the computation for the Model B is only possible by using the B-spline panels as demonstrated by Kashiwagi (1998a). However, Model C would not be possible even with the use of the B-spline panel method. In other words, we are facing a bottle-neck for the large storage requirement.

One way to overcome this difficulty is to use a modern iterative solver such as GMRES. In such a modern iterative solver, the system matrix need not be stored. Instead, we have to evaluate the result of the matrix-vector multiplication at each iterative step. In the Green's function method, it is possible to evaluate the result of the matrix-vector multiplication without storing the system matrix itself. This reduces considerably the storage requirement to $O(N)$, and thus the bottle-neck for the storage requirement is solved.

Since the matrix-vector multiplication needs $O(N^2)$ computation time, the computational time would be the subsequent bottle-neck to be solved. In order to accelerate the computational time to $O(N \log N)$ or $O(N)$, the pre-corrected FFT method has been applied to MOB (Mobile Offshore Bases) by Korsmeyer et al. (1999) and Kring et al. (2000). On the other hand, the fast multipole method has been applied to a pontoon-type VLFS by Utsunomiya et al. (2001, 2006, 2007). The details of the methods are omitted here because of the advanced nature of these topics.

An alternative approach to tackle the hydroelastic problem for a "very large" floating structure is to use the finite element method (FEM) for solving the fluid domain as demonstrated by Seto et al. (1998, 2005). The advantage of FEM over BEM is that the resulting system matrix is a narrow-banded sparse matrix; thus it may be effectively solved as compared to a full matrix which appears in conventional BEM. FEM may also be effectively used for solving transient response of VLFS as shown by Watanabe et al. (1998) for a linear problem and by Kyoung et al. (2006) for a nonlinear problem.

3.3.7 Numerical examples

The wave response analysis of a box-like VLFS has been performed by using the fast multipole method (Utsunomiya et al. 2006). The VLFS specifications are: length $L = 1500$ m, $B = 150$ m, draft $d = 1$ m, the rigidity as an elastic plate $D = 3.88 \times 10^7$ kNm, and Poisson's ratio $v = 0.3$. The modal functions employed total 160 (20 in longitudinal and 8 in beam; in this numerical example, the products of free-free beam modes in air are used as the modal functions instead of using dry modes modeled by FEM). The wave period is 18 s (the corresponding wave length at the water depth of 8 m is $\lambda = 156.8$ m), and the angle of wave incidence is $\beta = \pi/4$.

Figure 3.7 shows the contour plot of the variable depth configuration, over which the VLFS is floated. The water depth at infinity is assumed to be 100 m. This configuration corresponds to the experiment conducted at the Port and Airport Research Institute in Japan (Shiraishi et al. 2001). The variable depth surface (S_B) is discretized into 20,278 elements of 12.5 m × 12.5 m in size; nodes totalling 61,721. For meshing the VLFS model, the same element size is used (nodes total 5,377). Thus, the total node number of the analyzed model becomes 67,098. The computation time was 38.4 h using 5 CPUs for the 18 s wave period, with the residual tolerance $\varepsilon = 10^{-2}$ in the GMRES solver.

Figure 3.8 shows the deflection amplitudes in the variable depth sea and in the constant depth sea ($h = 8$ m). We observe a considerable difference in response characteristics when the VLFS is in the variable depth sea or the

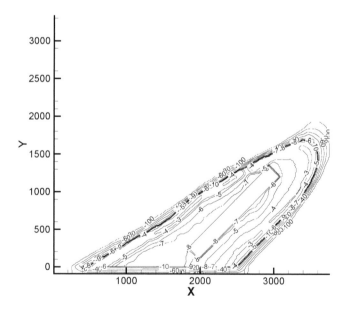

Figure 3.7 Contour plot of variable depth configuration.

Source: Utsunomiya et al. (2006).

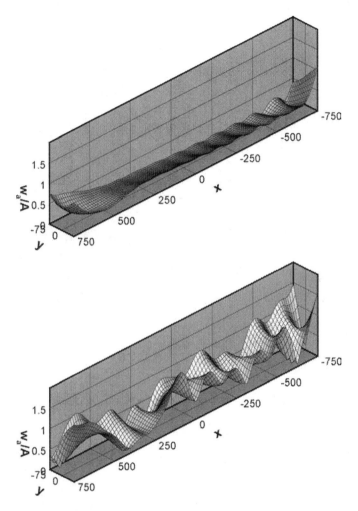

Figure 3.8 Deflection amplitude at $T = 18$ s.
Source: Utsunomiya et al. (2006).

constant depth sea. Thus, in a relatively shallow water case, the effect of variable depth under the VLFS is significant and must be accounted for.

Figure 3.9 shows a snapshot of the surface elevation around the floating body, where both diffraction and radiation waves are included. One can clearly see the refraction of waves, and the change of wave length (about $\lambda = 157$ m) inside the reef, corresponding to 8 m water depth) by dispersion in Figure 3.9.

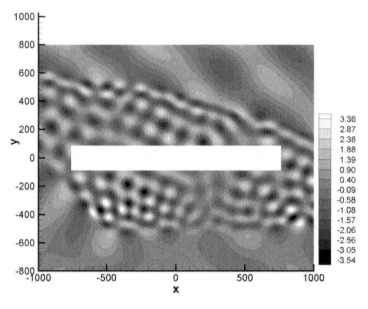

Figure 3.9 Snapshot of surface elevation around floating body in variable depth sea.
Source: Utsunomiya et al. (2006).

3.4 Conclusions

In this chapter, hydroelastic analysis of VLFS was introduced, where a pontoon type VLFS was mainly focused. As an example of the analytical approach, the eigenfunction expansion matching method for a floating plate modeled as a 2-D problem was presented. Then, the numerical approach using dry-mode superposition method and the Green's function method was explained. The dry-mode superposition approach is most commonly used for predicting the hydroelastic behavior. Several methods for acceleration of the Green's function method to solve "very large" hydroelastic problems were also briefly introduced.

References

Eatock Taylor, R., and Waite, J.B. (1978) "Dynamics of offshore structures evaluated by boundary integral techniques," *International Journal for Numerical Methods in Engineering*, 13, pp. 73–92.

Endo, H. (2000) "The behavior of a VLFS and an airplane during takeoff/landing run in wave condition," *Marine Structures*, 13, pp. 477–491.

Fujikubo, M. (2005) "Structural analysis for the design of VLFS," *Marine Structures*, 18, pp. 201–226.

Iijima, K., Yoshida, K., and Suzuki, H. (1998) "Hydroelastic analysis of semi-submersible type VLFS capable of detailed structural analysis," *Hydroelasticity in Marine Technology*, Kashiwagi, M., Koterayama, W., and Ohkusu, M. (eds), RIAM, Kyushu University, pp. 211–218.

Inoue, K., Nagata, S., and Niizato, H. (2003) "Stress analysis of detailed structures of Mega-Float in irregular waves using entire and local structural models," *Proceedings of the 4th International Workshop on Very Large Floating Structures*, Tokyo, Japan, pp. 219–228.

John, F. (1950) "On the motion of floating bodies II," *Communications on Pure and Applied Mathematics*, 3, pp. 45–101.

Kashiwagi, M. (1998a) "A B-spline Galerkin scheme for calculating the hydroelastic response of a very large floating structure in waves," *Journal of Marine Science and Technology*, 3, pp. 37–49.

Kashiwagi, M. (1998b) "A new solution method for hydroelastic problems of a very large floating structure in waves," *Proceedings of 17th International Conference on Offshore Mechanics and Arctic Engineering*, Lisbon, Portugal, OMAE98-4332.

Kashiwagi, M. (2000a) "Research on hydroelastic responses of VLFS: Recent progress and future work," *International Journal of Offshore and Polar Engineering*, 10, pp. 81–90.

Kashiwagi, M. (2000b) "A time-domain mode-expansion method for calculating transient elastic responses of a pontoon-type VLFS," *Journal of Marine Science and Technology*, 5, pp. 89–100.

Kim, J.W. and Webster, W.C. (1998) "The drag on an airplane taking off from a floating runway," *Journal of Marine Science and Technology*, 3, pp. 76–81.

Korsmeyer, T., Klemas, T., White, J., and Phillips, J. (1999) "Fast hydrodynamic analysis of large offshore structures," *Proceedings of 9th International Offshore and Polar Engineering Conference*, Brest, France, Vol. 1, pp. 27–34.

Kring, D., Korsmeyer, T., Singer, J., and White, J. (2000) "Analyzing mobile offshore bases using accelerated boundary-element method," *Marine Structures*, 13, pp. 301–313.

Kyoung, J.H., Hong, S.Y., and Kim, B.W. (2006) "FEM for time domain analysis of hydroelastic response of VLFS with fully nonlinear free-surface conditions," *International Journal of Offshore and Polar Engineering*, 16, pp. 168–174.

Mamidipudi, P. and Webster, W.C. (1994) "The motions performance of a mat-like floating airport," *Hydroelasticity in Marine Technology*, Faltinsen, O., Larsen, C.M., Moan, T., Holden, K., and Spisdoe, N. (eds), A.A. Balkema, Rotterdam, pp. 363–375.

McIver, M. (1985) "Diffraction of water waves by a moored, horizontal, flat plate," *Journal of Engineering Mathematics*, 19, pp. 297–319.

Nagata, S., Yoshida, H., Fujita, T., and Isshiki, H. (1998) "The analysis of wave-induced response of an elastic floating plate in a sea with a breakwater," *Proceedings of 8th International Offshore and Polar Engineering Conference*, Montreal, Canada, Vol. I, pp. 223–230.

Newman, J.N. (1985) "Algorithms for the free-surface Green function," *Journal of Engineering Mathematics*, 19, pp. 57–67.

Newman, J.N. (1994) "Wave effect on deformable bodies," *Applied Ocean Research*, 16, pp. 47–59.

Newman, J.N. (2005) "Efficient hydrodynamic analysis of very large floating structures," *Marine Structures*, 18, pp. 169–180.

Ohkusu, M. and Nanba, Y. (1996) "Analysis of hydroelastic behavior of a large floating platform of thin plate configuration in waves," *Proceedings of International Workshop on Very Large Floating Structures*, Hayama, Japan, pp. 143–148.

Ohmatsu, S. (1998) "Numerical calculation of hydroelastic behavior of pontoon type VLFS in waves," *Proceedings of 17th International Conference on Offshore Mechanics and Arctic Engineering*, Lisbon, Portugal, OMAE98-4333.

Ohmatsu, S. (2005) "Overview: Research on wave loading and responses of VLFS," *Marine Structures*, 18, pp. 149–168.

Ohta, H., Torii, T., Hayashi, N., Watanabe, E., Utsunomiya, T., Sekita, K., and Sunahara, S. (1998) "Effect of attachment of horizontal/vertical plate on the wave response of a VLFS," *Proceedings of 3rd International Workshop on Very Large Floating Structures*, Ertekin, R.C. and Kim, J.W. (eds), University of Hawaii at Manoa, USA, pp. 265–274.

Seto, H. and Ochi, M. (1998) "A hybrid element approach to hydroelastic behavior of a very large floating structure in regular waves," *Hydroelasticity in Marine Technology*, Kashiwagi, M., Koterayama, W., and Ohkusu (eds), RIAM, Kyushu University, pp. 185–193.

Seto, H., Ohta, M., Ochi, M., and Kawakubo, S. (2005) "Integrated hydrodynamic-structural analysis of very large floating structures (VLFS)," *Marine Structures*, 18, pp. 181–200.

Shiraishi, S., Harasaki, K., Yoneyama, H., Iijima, K., and Hiraishi, T. (2001) "Experimental study on elastic response and mooring forces of very large floating structures moored inside reef," *Proceedings of 20th International Conference on Offshore Mechanics and Arctic Engineering*, Rio de Janeiro, Brazil, OMAE01-5203.

Suzuki, H. et al. (2006) "ISSC committee VI.2: Very Large Floating Structures," *16th International Ship and Offshore Structures Congress*, Southampton, UK, pp. 391–442.

Takagi, K., Shimada, K., and Ikebuchi, T. (2000) "An anti-motion device for a very large floating structure," *Marine Structures*, 13, pp. 421–436.

Teng, B. and Eatock Taylor, R. (1995) "New higher-order boundary element methods for wave diffraction/radiation," *Applied Ocean Research*, 17, pp. 71–77.

Utsunomiya, T. and Okafuji, T. (2007) "Wave response analysis of a VLFS by accelerated Green's function method in infinite water depth," *International Journal of Offshore and Polar Engineering*, 17, pp. 30–38.

Utsunomiya, T. and Watanabe, E. (2006) "Fast multipole method for wave diffraction/radiation problems and its application to VLFS," *International Journal of Offshore and Polar Engineering*, 16, pp. 253–260.

Utsunomiya, T., Watanabe, E., and Eatock Taylor, R. (1998) "Wave response analysis of a box-like VLFS close to a breakwater," *Proceedings of 17th International Conference on Offshore Mechanics and Arctic Engineering*, Lisbon, Portugal, OMAE98-4331.

Utsunomiya, T., Watanabe, E., and Nishimura, N. (2001) "Fast multipole algorithm for wave diffraction/radiation problems and its application to VLFS in variable water depth and topography," *Proceedings of 20th International Conference on Offshore Mechanics and Arctic Engineering*, Rio de Janeiro, Brazil, OMAE01-5202.

Watanabe, E. and Utsunomiya, T. (1996) "Transient response analysis of a VLFS at airplane landing," *Proceedings of International Workshop on Very Large Floating Structures*, Hayama, Japan, pp. 243–247.

Watanabe, E., Utsunomiya, T., and Tanigaki, S. (1998) "A transient response analysis of a very large floating structure by FEM," *Structural Engineering/Earthquake Engineering*, JSCE, 15, pp. 155–163.

Watanabe, E., Utsunomiya, T., and Wang, C.M. (2004a) "Hydroelastic analysis of pontoon-type VLFS: A literature survey," *Engineering Structures*, 26, pp. 245–256.

Watanabe, E., Wang, C.M., Utsunomiya, T., and Moan, T. (2004b) "Very large floating structures: Applications, analysis and design," *CORE Report 2004-02*, National University of Singapore, www.eng.nus.edu.sg/core/publicationsresearchreports.htm

Watanabe, E., Utsunomiya, T., Kuramoto, M., Ohta, H., Torii, T., and Hayashi, N. (2003) "Wave response analysis of VLFS with an attached submerged plate," *International Journal of Offshore and Polar Engineering*, 13, pp. 190–197.

Wu, C., Watanabe, E., and Utsunomiya, T. (1995) "An eigenfunction matching method for analyzing the wave induced responses of an elastic floating plate," *Applied Ocean Research*, 17, pp. 301–310.

Yago, K. and Endo, H. (1996) "On the hydroelastic response of box-shaped floating structure with shallow draft," *Journal of the Society of Naval Architects of Japan*, 180, pp. 341–352 (in Japanese).

Chapter 4

Structural analysis and design of VLFS

Masahiko Fujikubo

4.1 Introduction

The objective of structural design is to develop a structure that fulfils serviceability and safety requirements in a cost-effective manner. For a novel structure like a very large floating structure (VLFS), the structural design needs a first-principle approach that is based on a rational structural response analysis and explicit design criteria.

Hydroelastic response analysis is an essential process for predicting the global response of deformation and stresses in VLFSs. In the initial stage of the structural design of a VLFS, a uniform plate model is generally used for the hydroelastic response analysis. Such a simple structural model is effective for the determination of global stiffness parameters to meet design requirements. For the subsequent design stage, however, a more detailed structural model that can deal with structural member behaviors is needed in order to determine the structural member arrangement and configuration. Owing to its huge structural size, the applicability of three-dimensional (3D) finite element (FE) modeling to the entire VLFS is rather limited. Therefore, simple and rational structural modeling techniques need to be developed. Furthermore, hydroelastic response analysis should ultimately furnish the local stress response of detailed structures, as well as global responses. A hierarchical system of structural analysis that includes a zooming technique is needed.

The design limit states for marine structures include the ultimate limit state (ULS), the fatigue limit state (FLS), the serviceability limit state (SLS), and the progressive collapse limit state (PLS). ULS refers to the ultimate event in which structural resistance has an appropriate reserve. PLS refers to the progressive failure of structures when subjected to accidental or abnormal load effects (DNV 1997). PLS for accidental load effects is also called the accidental collapse limit state (ALS) (Moan 2004). ULS corresponds to component design verification based on the elastic behavior of the structure, and thus can be examined by the hydroelastic response analysis. But PLS needs a progressive collapse analysis that considers nonlinear structural behaviors such as buckling and yielding.

In this chapter, the linear and nonlinear structural analysis methods for the design of VLFSs are described, based on the achievements made on the pontoon-type VLFS during and after the Mega-Float project in Japan. First, a structural design flow, typical of a VLFS, and the design limit states specified in the Technical Guidelines of Mega-Float, are explained to clarify the role of linear and nonlinear structural analyses at various stages of the structural design. Structural modeling techniques developed for the hydroelastic response analysis of a pontoon-type VLFS and an effective method for predicting the local stress response using a multi-step approach are explained next. Regarding nonlinear structural analysis, example analyses of structural damage due to airplane collision are presented. Finally, a simplified method of progressive collapse analysis of a pontoon-type VLFS using the idealized structural unit method (ISUM) is presented as an effective tool for overall collapse analysis of VLFSs.

4.2 Structural design of a VLFS

4.2.1 Design flow

In the case of conventional ships and floating offshore structures, the global response is dominated by rigid-body motions, whereas in VLFSs, the global response is determined by elastic responses, either dynamically or statically. Structural stiffness is therefore a governing parameter for the design of a VLFS. The relationship between structural stiffness and global elastic responses is generally complex as a result of fluid–structure interaction effects. For example, an increase in structural stiffness can lead to an increase in cross-sectional forces and stresses. Accordingly, hydroelastic response analysis must be performed at every structural design stage in order to consider the effects of design changes on structural stiffness and responses. Figure 4.1 shows a typical design flow for a VLFS having such characteristics, generated by the design practices of Mega-Float (ISSC 2006). As shown, the design flow can be divided into three basic stages. During the first stage, a relatively simple method of hydroelastic response analysis is used, and the fundamental magnitude of cross-sectional stiffness, and the corresponding basic design variables, such as structural depth, primary-members' arrangement, and size are determined. The characteristic length and frequency, which were derived by Suzuki (1996) as a dominant design parameter with respect to hydroelastic responses of VLFSs, are referenced in the process of determining the global stiffness. The hydroelastic response analyses that assume the uniform rectangular plate model (e.g. Kashiwagi 1996; Ohmatsu 1997) are generally employed during this stage. A combination of an FE plate model and a modal approach (Okada 1999) is also applied when a more refined structural modeling (e.g. a variable flexural stiffness) is needed.

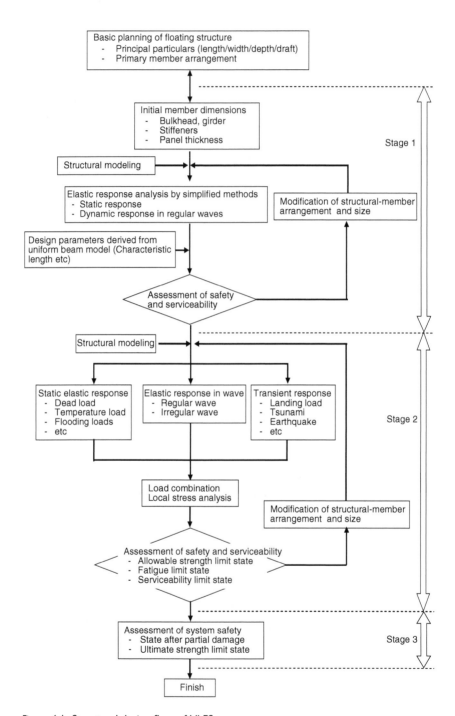

Figure 4.1 Structural design flow of VLFS.

In the second stage, the detailed design for actual structural configurations involving variable structural depth, variable planar shape, openings in bulkheads for the usage of internal space, and so on is performed. The 3D detail method of hydroelastic response analysis, developed by Seto (2003), is generally applied at this stage. The fluid model, implemented in this method, can address an uneven body boundary, a variable sea depth, and the presence of breakwaters and seashore. From the results of the global response analysis, the local stress response under combined load effects is evaluated using a zooming technique. Through the evaluation of strength and SLSs, the size and arrangement of structural members are determined.

In Figure 4.1, both the initial (first) and detail (second) design stages have a design loop that includes the hydroelastic global response analysis. This is a typical feature of the structural design of a VLFS. In this chapter, structural modeling techniques that are employed for global elastic response analysis during the second design stage are explained in Section 4.3.1. With regard to the local strength analysis, a method of local stress analysis of detailed structure in irregular waves, proposed by Inoue (2003), is introduced in Section 4.3.2.

During the third stage of the design flow, the structural safety assessments for damaged conditions and for the collapse behavior under abnormal load effects are performed. The safety assessment at this stage generally needs a nonlinear progressive collapse analysis. More details on the design limit states are provided in Section 4.2.2, and an example of progressive collapse analyses is given in Section 4.4.

4.2.2   Design limit states

The Technical Guidelines for Mega-Float specifies four design limit states with respect to structural safety, as shown in Table 4.1 (TRAM 1999). The allowable strength limit state is intended to confirm that a structure has strength for safe use throughout the service period with an appropriate reserve. Buckling and yielding strengths of principal structural members are evaluated based on their elastic response. The Class-1 environmental load that corresponds to a return period of at least two times the number of service year is considered as a characteristic value. The ultimate strength limit state is intended to confirm that a structure, under abnormal load effects from waves, earthquakes and tsunamis, does not achieve a full collapse state, where structural damage may lead to a loss of human life, sinking or drifting. A Class-2 environmental load that corresponds to a return period of at least two times the number of service year is considered as a characteristic value. The state that occurs after partial damage refers to the progressive collapse limit state under accidental load effects. The possibility of load redistribution without any progressive collapse is examined for the

Table 4.1 Design limit states for structural safety

Limit states	Definition	Characteristic value of environmental loads	
		Case A	Case B
Allowable strength limit state	A state in which a structure or principal structural member reaches strength limit state, where a slight damage that will not hinder the use of the structure may occur	Class 1	Class 1
Ultimate strength limit state	A state in which a structure or a principal structural member reaches a full collapse state, where a structural damage dangerous to human lives, a sinking or a drifting occurs	Class 2	Class 2
Fatigue limit state	A state in which cracks initiate and propagate due to fatigue caused by a repeated load and the load bearing capacity of the member starts to decrease	Expected load frequency	
State after partial damage	A state in which load redistribution due to some damage makes the other intact structural members or structure damaged	Class 0	

Notes
Case A: Wave, wind, current and others excluding Case B; Case B: earthquake, seaquake, tsunami and storm surge; Class 0: return period of two years; Class 1: return period of at least two times service year; Class 2: return period of at least 100 times service years.

Class-0 environmental loads that correspond to a return period of at least two years.

In addition to these strength limit states, SLSs are considered. The Class-0 environmental load with a return period of at least two years is used as a characteristic value. A typical example of SLS criteria for VLFS, when used for a floating airport, is the stringent tolerance in the radius of the curvature for the runway part. According to the airport facility design standard, the radius of curvature should be larger than 30,000 m. This sometimes gives a more severe design condition than the strength limit states. It is a feature of the design of VLFS that SLS criteria with respect to top-side structures are generally severe compared to those for conventional ships and offshore floating structures.

In the design limit states, the allowable strength, fatigue, and serviceability limit states are examined by elastic response analyses, while the ultimate strength limit state and the state after damage are examined by nonlinear progressive collapse analyses.

4.3 Elastic response analysis

4.3.1 *Structural modeling for global response analysis*

Figure 4.2(a) shows a typical structural arrangement for a pontoon-type VLFS. Similar structures may be found in the deck part of a semi-submersible VLFS. The structure consists of a deck, a bottom, and orthogonally intersecting bulkheads. Figure 4.2(b) shows a bulkhead with openings for the usage of internal spaces, such as for a transportation system. The most accurate way to model these structures is clearly a 3D-shell FE model as shown in Figure 4.2(b). For the entire VLFS, however, the degrees of freedom become excessively large. A more simplified structural modeling is generally needed for global response analysis. They include an orthotropic plate model, a plane grillage model, and a sandwich-grillage model as described below.

Orthotropic plate model

The intersection part of longitudinal and transverse bulkheads (see Figure 4.3(a)) is a unit structure of a VLFS. The orthotropic plate model, as shown in Figure 4.3(b), idealizes the unit structure as a uniform plate element having equivalent bending stiffness and torsional stiffness. This

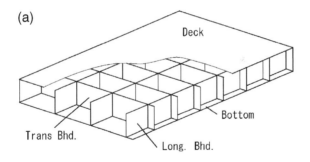

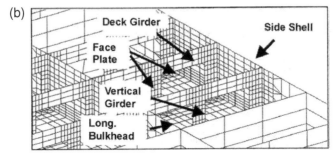

Figure 4.2 Structural arrangement of a Pontoon-type VLFS. (a) Typical arrangement of primary members. (b) Bulkhead with opening (3D shell model).

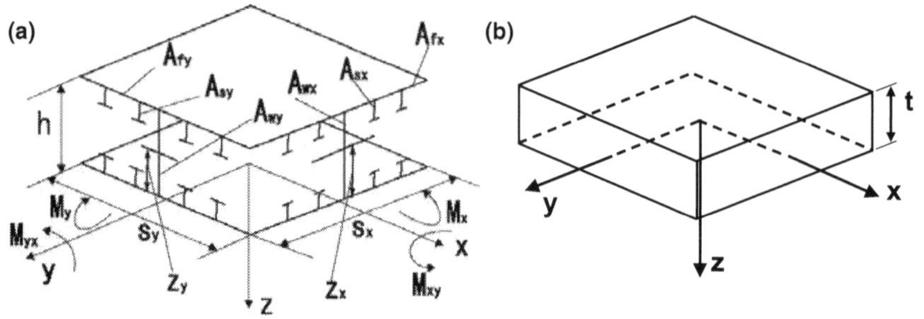

Figure 4.3 Transformation of actual structure to equivalent orthotropic plate. (a) Unit structure. (b) Orthotropic plate model.

transformation to the orthotropic plate model has a wide application to either the semi-analytical approach for a uniform rectangular plate model (Kashiwagi 1996; Ohmatsu 1997) or the full numerical approach based on the FE model (Okada 1999; Seto 2003). The material properties for the equivalent orthotropic plate model can be determined as shown below.

By denoting the global deflection of structure by $w(x, y)$, the biaxial bending moments per unit width, M_x and M_y, and the torsional moment per unit width, M_{xy}, are given in the form

$$\begin{Bmatrix} M_x \\ M_x \\ M_{xy} \end{Bmatrix} = \begin{bmatrix} D_x & D_{xy} & 0 \\ D_{yx} & D_y & 0 \\ 0 & 0 & G_{xy} \end{bmatrix} \begin{Bmatrix} -\dfrac{\partial^2 w}{\partial x^2} \\ -\dfrac{\partial^2 w}{\partial y^2} \\ -2\dfrac{\partial^2 w}{\partial x \partial y} \end{Bmatrix} \tag{4.1}$$

where the coefficients $D_x, D_y, D_{xy} \, (=D_{yx})$, and G_{xy} for the actual structure are given by the simple equations as derived in the appendix (Fujikubo 2005).

On the other hand, the stress–strain relationship of orthotropic material under plane-stress conditions is expressed by

$$\begin{Bmatrix} \sigma_x \\ \sigma_y \\ \tau_{xy} \end{Bmatrix} = \begin{bmatrix} \dfrac{E_1}{1 - v_{12}v_{21}} & \dfrac{v_{21}E_1}{1 - v_{12}v_{21}} & 0 \\ \dfrac{v_{12}E_2}{1 - v_{12}v_{21}} & \dfrac{E_2}{1 - v_{12}v_{21}} & 0 \\ 0 & 0 & G_{12} \end{bmatrix} \begin{Bmatrix} \varepsilon_x \\ \varepsilon_y \\ \gamma_{xy} \end{Bmatrix} \tag{4.2}$$

where $v_{21}E_1 = v_{12}E_2$. The moment–curvature relationship of the equivalent orthotropic plate model of the same material is given by

$$\left\{ \begin{array}{c} M_x \\ M_y \\ M_{xy} \end{array} \right\} = \frac{t^3}{12} \left[\begin{array}{ccc} \dfrac{E_1}{1 - v_{12}v_{21}} & \dfrac{v_{21}E_1}{1 - v_{12}v_{21}} & 0 \\ \dfrac{v_{12}E_2}{1 - v_{12}v_{21}} & \dfrac{E_2}{1 - v_{12}v_{21}} & 0 \\ 0 & 0 & G_{12} \end{array} \right] \left\{ \begin{array}{c} -\dfrac{\partial^2 w}{\partial x^2} \\ -\dfrac{\partial^2 w}{\partial y^2} \\ -2\dfrac{\partial^2 w}{\partial x \partial y} \end{array} \right\} \qquad (4.3)$$

where t is the thickness of the equivalent orthotropic plate. In view of the relationship $v_{21}E_1 = v_{12}E_2$ and by taking an arbitrary value as the thickness t, there are four independent variables, namely E_1, E_2, v_{12}, and G_{12} in Eq. (4.3). By equating D_x, D_y, D_{xy} ($=D_{yx}$), and G_{xy} in Eq. (4.1) and the corresponding components in Eq. (4.3), we obtain four equations that allows us to determine the material properties E_1, E_2, v_{12}, and G_{12}.

Plane grillage model

In the plane grillage model, the structure (see Figure 4.3(a)) is modeled by beam elements as shown in Figure 4.4(a). The deck and bottom panels are included in the beam elements as their effective faceplate. Since the continuity of deck and bottom plating is not modeled, the Poisson's effect is neglected. With regard to the transverse shear deformation of deck and bottom plating under the action of torsional moment, however, the equivalent torsional stiffness for beam elements can be formulated by considering the characteristics of torsional deformation of the actual, continuous structure and based on a concept of equivalent torsional strain energy (Fujikubo 2001). The effect of opening, as shown in Figure 4.2, on the shear stiffness of bulkheads can be easily taken into account by adjusting the beam stiffness.

A major drawback of this model is the neglect of Poisson's effect. Figure 4.5 shows the response amplitude of a pontoon-type VLFS

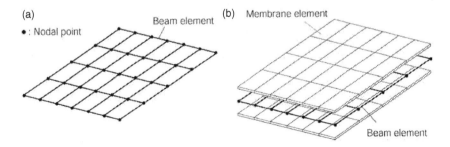

Figure 4.4 Grillage models. (a) Plane grillage model. (b) Sandwich grillage model.

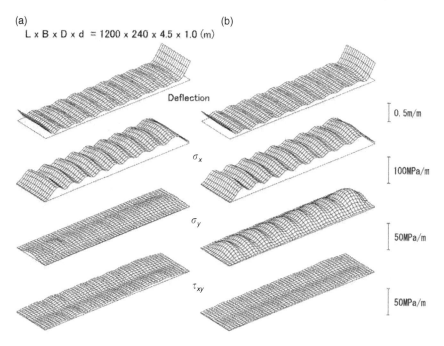

(a)   (b)

L x B x D x d = 1200 x 240 x 4.5 x 1.0 (m)

Deflection

σ_x

σ_y

τ_{xy}

0. 5m/m

100MPa/m

50MPa/m

50MPa/m

Figure 4.5 Response amplitudes distribution in longitudinal regular wave. (a) Plane grillage model. (b) Sandwich grillage model.

Source: Fujikubo (2001).

Note
Wavelength to structural length ratio = 0.08.

($L \times B \times D \times d = 1,200$ m $\times$ 240 m $\times$ 4.5 m $\times$ 1.0 m) in a longitudinal regular wave. For a VLFS having a mat-like shape, the Poisson's effect has a significant influence on the prediction of global response, particularly stress responses. In this wave condition, the region near the longitudinal centerline is nearly in a plane-strain state, owing to the Poisson's effect, and the transverse bending stress is significantly induced, as well as the longitudinal one. Whilst this response can be captured by the plate model and the sandwich grillage model (see Figure 4.5(b)), it cannot be captured by the plane-grillage model.

Sandwich grillage model

In the sandwich grillage model, the deck and bottom panels are modeled by the rectangular membrane elements, while the bulkheads by the beam elements as shown in Figure 4.4(b). The Poisson's effect and shear deformation at the deck and bottom plating can be automatically considered in the membrane elements. Based on the Kirchhoff–Love assumption that the cross-section remains plane and perpendicular to the neutral plane, the inplane degrees of freedom of the membrane elements are expressed as a function of nodal displacements of beam elements (Fujikubo 2001). The total degrees of freedom are the same as those for the plane grillage model.

The sandwich grillage model accounts for the bending, torsional and vertical shear stiffnesses in a straightforward manner and yet the computational efficiency is similar to that of the plane grillage model. This is therefore the most effective model for the global response analysis of a VLFS.

4.3.2 Stress analysis of local structures

In order to predict the stress response of local structures including the structural details necessary for fatigue strength assessment, a zooming analysis based on the results of global response analysis has to be performed. A VLFS in multi-directional wave conditions is subjected to combined load effects due to biaxial bending moments, shear forces, and torsional moments. It requires tremendous computational efforts to perform the stress analysis of local structures for all possible load combinations in multi-directional irregular waves.

Inoue (2003) proposed an efficient method, called the stress factor method, for predicting the stress response of local structures of a VLFS in multi-directional irregular waves. Figure 4.6 shows the basic flow of the stress factor method. First, the hydroelastic global response analysis for regular waves with specified period T and incident angle θ is performed to obtain the complex amplitude $P_i(\theta, T)$ of the load components i, which includes the internal load effects (e.g. bending moments and shear forces) as well as the external load components (e.g. hydrodynamic pressure and inertia force). FE analysis of local structure is performed for the unit magnitude of each load component i. The resulting local stress distribution is defined as a stress factor $\sigma_{\mathrm{P}i}$ for load component i. The complex stress amplitudes of local structures under load component i are calculated by multiplying the stress factor $\sigma_{\mathrm{P}i}$ and the complex amplitude of load component $P_i(\theta, T)$. The complex stress amplitudes under the combined action of load components in the specified regular wave condition (θ, T) is then calculated as

$$\sigma(\theta, T) = \sum_i P_i(\theta, T)\sigma_{\mathrm{P}i} \tag{4.4}$$

The response amplitude operator (RAO) is calculated as shown in Figure 4.6. Using the RAOs, the stochastic predictions of local stress response are made. Once the stress factors for a unit magnitude of each load component are obtained by the local FE analysis, the RAO of the local stresses can be estimated from the result of the hydroelastic global response analysis in an efficient manner. An essential point to be noted in the stress factor method is

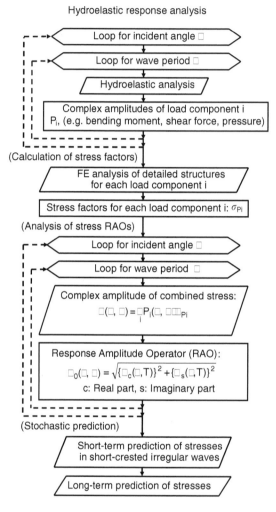

Figure 4.6 Flow of stress analysis in irregular waves using stress factors for local structural models.

Source: Inoue (2003).

that the load and boundary conditions for the FE analysis of local structures must be carefully selected so that the other load components do not affect the results.

Figure 4.7 shows the local structural model for the analysis of stress response at the slots of bottom longitudinals (Inoue 2003). The structure is subjected to hydrodynamic pressures on the bottom shell, transverse

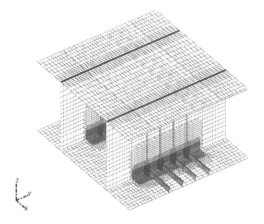

Figure 4.7 Local structure with slots of bottom longitudinals.
Source: Inoue (2003).

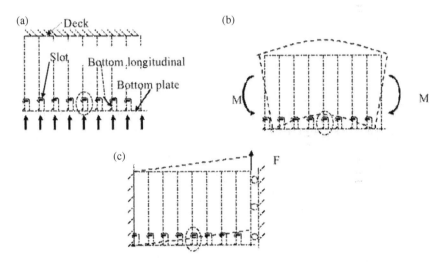

Figure 4.8 Local model to calculate stress factors of slots of bottom longitudinals. (a) For the dynamic pressure. (b) For the transverse bending moment. (c) For the transverse shear force.
Source: Inoue (2003).

bending moments and transverse shear forces in the transverse web. The boundary conditions and the unit load for the analysis of the stress factor with respect to each load component are shown in Figure 4.8, and the calculated stress factors for the case of shear force in Figure 4.9. The applicability of the stress factor method was demonstrated through a comparison

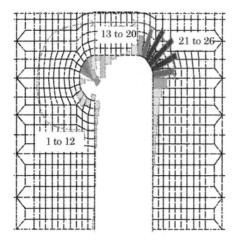

Figure 4.9 Stress factors for shear force along the edge of a slot of a bottom longitudinal.

Source: Inoue (2003).

of measured and estimated significant stress values of a pontoon-type VLFS ($L \times B \times D \times d = 200$ m $\times$ 100 m $\times$ 3 m $\times$ 1.0 m), located in Tokyo-Bay, under short-term wave conditions (Oka 2003).

4.4   Collapse analysis

4.4.1   Damage analysis during airplane collision

A VLFS may experience accidental damages due to airplane collision, ship collision, fire, and explosion during its operation. As an example of a PLS check of a VLFS, damage analysis of a pontoon-type VLFS during an airplane collision is presented (TRAM 1997; Muragishi 1997).

The collision of a 500-t airplane (B747) with the runway of the floating airport was studied for the following three collision scenarios:

1   Vertical drop on deck
2   Collision to sidewall due to short landing
3   Belly landing on deck

Figure 4.10 shows the FE model for the analysis of a vertical drop on deck. The runway is 7-m deep and paved with pre-stressed concrete of 300-mm thickness. The longitudinal girders and transverse frames are provided at an interval of 30 m and 6.55 m, respectively. The area extending

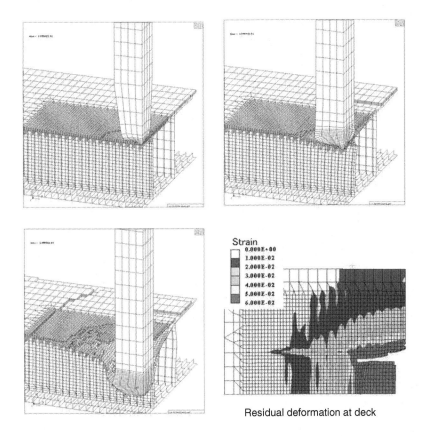

Residual deformation at deck

Figure 4.10 Collision deformation due to vertical drop.

Source: TRAM (1997).

Note
Collision velocity = 100 m/s.

over three longitudinal girder spacing and five transverse frame spacing was used for the analysis, and the collision point was located at the center of the area. Based on the symmetry of the model, a quarter of the model was analyzed. The preliminary global response analysis for a larger area that used a simple mass and nonlinear-spring model for the collision point showed that the area shown in Figure 4.10 was as large as the possible largest damaged area. On this basis, the outer edge of the model was fixed. Since the global response velocity of the structure was small enough compared to the collision velocity, the change of the buoyancy forces during collision was neglected.

The nonlinear explicit FE code, LS-DYNA3D, was employed for the analysis. The floating structure was modeled by shell and beam elements and the

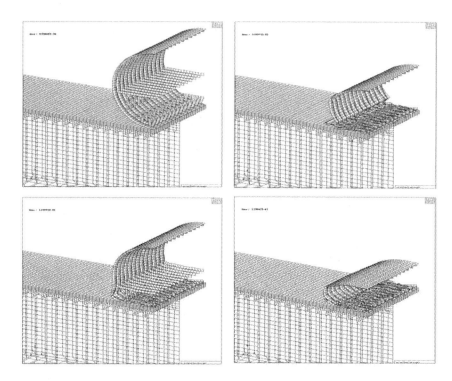

Figure 4.11 Collision deformation due to belly landing.

Source: TRAM (1997).

Note
Collision velocity = 100 m/s.

concrete pavement by solid elements. The airplane was idealized by solid elements having the equivalent axial load-shortening relationship and the mean crash strength. Figure 4.10 shows the collision deformation for the collision velocity of 100 m/s. For the velocity of 70 m/s, no penetration of the deck's steel panel was observed, whereas for the velocity of 100 m/s, deck penetration took place. The damage, however, did not lead to the penetration of the bottom. As a possible consequence of deck penetration, as shown in Figure 4.10, a simulation of the fire from fuel leakage in the holding space at the collision point was performed by Yoshida (2005), and the structural safety was assessed.

Figure 4.11 shows the collision deformation due to a belly landing. It can be seen that for a landing velocity of 100 m/s, the airplane is almost completely crushed although only little damage is observed in the floating structure.

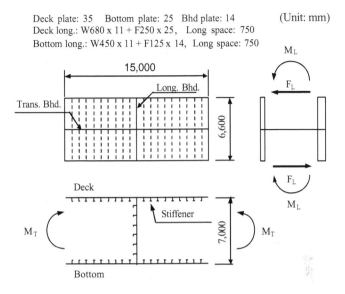

Deck plate: 35   Bottom plate: 25   Bhd plate: 14   (Unit: mm)
Deck long.: W680 x 11 + F250 x 25,   Long space: 750
Bottom long.: W450 x 11 + F125 x 14,   Long space: 750

Figure 4.12 Unit structure of VLFS.

Source: Fujikubo (2003a).

4.4.2   Collapse analysis of VLFS in waves

Method of nonlinear structural analysis

In order to assess the structural safety of VLFS under abnormal environmental conditions, it is important to determine the ultimate failure mode of an entire structure and the corresponding ultimate capacity, by performing a progressive collapse analysis. The unit structure of a VLFS having I-type cross-section, as shown in Figure 4.3, has little reserved capacity after either a deck panel or a bottom panel have reached their ultimate strength. The load redistribution after the collapse of primary structural member is therefore expected only in the planar directions. This again shows the importance of the progressive collapse analysis at an overall structural level.

As a result of the progress of computational capability, the nonlinear FE analysis is applied to the collapse analysis of structural systems, as exemplified in Figures 4.10 and 4.11. Such a direct FE simulation, however, still cannot be applied to the entire VLFS having huge dimensions.

A simplified method for the progressive collapse analysis of a VLFS using the ISUM is presented in this section. ISUM, in its original version developed by Ueda (1984), is based on a matrix formulation similar to conventional FE analysis, but it employs particular definitions of elements, which are of the same size and scale as the structural members themselves. Nonlinear buckling behavior of a structural unit (e.g. one plate panel between stiffeners) is idealized by appropriate shape functions assumed based on the collapse modes. Nonlinear contributions of post-buckling deflection to the membrane strains of plate panels are evaluated based on the theory of elastic, large deflection analysis. More details about ISUM formulation for the plate and stiffened plate models can be found in (Fujikubo 2002).

As a most fundamental case, the progressive collapse analysis of the unit structure taken from a floating airport VLFS model was performed, as shown in Figure 4.12 (Fujikubo 2003a). The deck and bottom stiffeners are modeled by beam-column elements, the plate panel between stiffeners by ISUM plate elements (one element for one bay), and the longitudinal and transverse bulkheads by Timoshenko beam elements. The number of elements is two orders of magnitude smaller than that of the conventional FE analysis. Figure 4.13 shows the calculated bending moment–curvature relationship for the case of longitudinal bending. The ultimate strength was attained when the deck or bottom panel on the compression side of bending reached their ultimate strength, and for further increase of applied curvatures the load-carrying capacity rapidly decreases.

Progressive collapse analysis of a pontoon-type VLFS in waves

By applying ISUM, the progressive collapse behavior of a pontoon-type VLFS in waves was analyzed (Fujikubo 2005). A floating airport model with two 4,000-m long runways, as shown in Figure 4.14, was taken as an

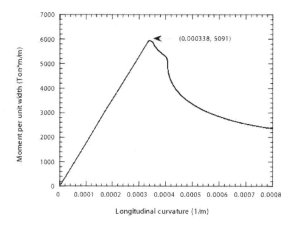

Figure 4.13 Moment–curvature relationships of unit structure of VLFS.

Source: Fujikubo 2003a.

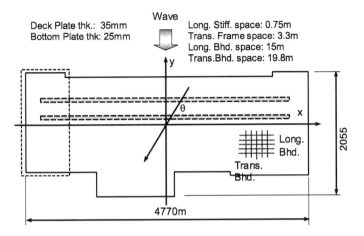

Figure 4.14 Model of prototype floating airport.

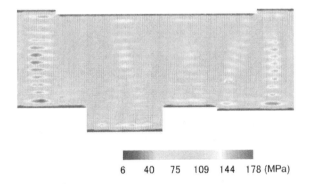

6   40   75   109   144   178 (MPa)

Figure 4.15 Extreme response of transverse stress at bottom in Class-2 irregular waves ($H_{1/3} = 4.8$ m, $T_{1/3} = 6.7$ sec).

example structure. Multi-directional (short-crested) irregular waves from the transverse direction were considered. This wave direction gives the severest design condition with respect to the buckling strength of deck and bottom plating having a longitudinal stiffening system (Fujikubo 2003a). The following three-step approach was applied.

Step 1: The hydroelastic global response analysis is performed. The equation of motion with respect to the nodal displacement $\{d\}$ can be expressed as

$$[M_S + M_A]\left\{\ddot{d}\right\} + [N]\left\{\dot{d}\right\} + \left[K_S^L + K_R\right]\{d\} = \{F_W\} \qquad (4.5)$$

with structural mass matrix $[M_S]$, added mass matrix $[M_A]$, hydrodynamic damping matrix $[N]$, linear elastic stiffness matrix of structure $[K_S^L]$, hydrostatic restoring force matrix $[K_R]$ and wave exciting force $\{F_w\}$.

Step 2: From the results of hydroelastic response analysis $\{d_e\}$, the external force vector $\{F_e\}$ including the inertia and damping effects is calculated from

$$\{F_e\} = \{F_W\} - [M_S + M_A]\left\{\ddot{d}_e\right\} - [N]\left\{\dot{d}_e\right\} \tag{4.6}$$

where $\{F_e\}$ gives the solution $\{d_e\}$ of Eq. (4.5), when it is applied to the floating structure in a quasi-static manner as

$$\left[K_S^L + K_R\right]\{d\} = \{F_e\} \tag{4.7}$$

Step 3: The time history of the external force vector $\{F_e\}$ is generated using the wave spectrum of irregular waves. It is then applied to the ISUM model in a quasi-static manner; that is the following nonlinear quasi-static equation is solved to obtain the progressive collapse behavior:

$$\left[K_S^{NL} + K_R\right]\{d\} = \{F_e\} \tag{4.8}$$

As described above, the progressive collapse analysis is performed by assuming that the dynamic external force distribution is the same as in the hydroelastic behavior and that only the structural stiffness changes as a result of the collapse. This assumption can be made when the collapsed area is limited as compared to the whole VLFS area (Fujikubo 2003b).

The Class-2 environmental condition for the ultimate strength limit state (see Table 4.1) was considered; that is, a 10,000-year return wave (100 times the number of service years) was estimated by the extreme wave statistics of Tokyo Bay. The significant wave height $H_{1/3}$ and wave period $T_{1/3}$ are 4.8 m and 6.7 s, respectively.

The hydroelastic global response analysis at Step 1 was performed using the 3D detail method (Seto 2003) and the equivalent orthotropic plate model. Figure 4.15 shows the extreme response of the transverse membrane stress at the bottom plating due to bending, corresponding to the probability of exceedance of 1/1,000 in the short-term sea state. The maximum stress response is observed on the lee side and near the transverse edges.

Using ISUM, the computational time for the progressive collapse analysis at Step 3 can be significantly reduced when compared to the conventional FE analysis. It was found, however, that the VLFS (see Figure 4.14) was still too large to be solved directly. So, considering the elastic stress response shown in Figure 4.15, a 1,650 m × 495-m area shown by the dashed lines in Figure 4.14 was analyzed. In addition, the collapse behavior of all

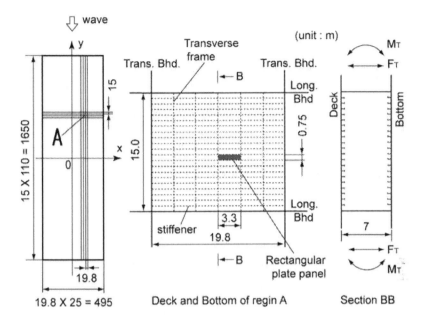

Figure 4.16 Model for collapse analysis.

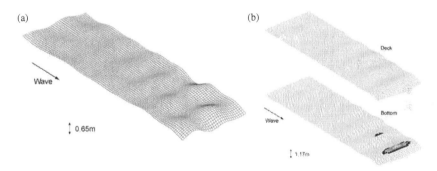

Figure 4.17 Deflection and spread of collapse region ($H_{1/3} = 4.8$ m, $T_{1/3} = 6.7$ sec). (a) Multi-directional irregular wave. (b) Uni-directional.

rectangular plate panels within the 19.8 m × 15 m area between two adjacent longitudinal and transverse bulkheads was assumed to be the same. The model for the ISUM analysis is shown in Figure 4.16. The sandwich grillage model was employed to allow for the shift in the neutral axis due to buckling and yielding of the deck and bottom panels.

Figure 4.17(a) shows the deflection mode when the maximum deflection took place in the generated 2-h time history of deflection in multi-directional irregular waves. None of the deck and bottom stiffened panels reached their ultimate strength in this case. For comparison purposes, the progressive collapse analysis in unidirectional irregular waves was performed, although it gives too severe a wave condition. In this case, the collapse occurred at the bottom panel with a smaller thickness than the deck panel as shown in Figure 4.17(b). It was found that the collapse area is confined to a relatively small area even in such an unrealistically severe wave condition.

The progressive collapse analysis, as shown above, is important not only for the prediction of the ultimate capacity of structure but also for the assessment of catastrophic failure scenario, its consequence, and the resulting risk.

4.5   Concluding remarks

Recent progress in linear and nonlinear structural analyses for the design of VLFSs has been presented. The major achievements can be summarized as a structural modeling technique that permits a rational analysis of VLFSs of huge structural sizes and a combination of the hydroelastic global response analysis with either the stress analysis of a detailed structure or the progressive collapse analysis of a global structure. Although not addressed in this chapter, significant progress has also been made in the probabilistic safety assessment of VLFS (Katsura 1999; Fujikubo 2003a).

Hitherto, studies on the structural analysis of VLFSs have been mainly concerned with a pontoon-type VLFS. More studies on structural modeling and progressive collapse analysis for semi-submersible-type VLFSs should be undertaken. There have been some attempts made on the structural optimization of a pontoon-type VLFS (Ma 1999; Yasuzawa 2003). The study on this area should be further pursued.

Appendix: Moment–curvature relationship of VLFS unit structure

The coordinate system whose origin is located on the neutral plane of the bottom plate panel is defined as shown in Figure 4.3(a). The stress–strain relationship of deck and bottom plate panels is given by

$$\begin{Bmatrix} \sigma_x \\ \sigma_y \\ \tau_{xy} \end{Bmatrix} = \frac{E}{1 - v^2} \begin{bmatrix} 1 & v & 0 \\ v & 1 & 0 \\ 0 & 0 & 1 - v/2 \end{bmatrix} \begin{Bmatrix} \varepsilon_x \\ \varepsilon_y \\ \gamma_{xy} \end{Bmatrix} \quad \text{(A.1)}$$

where E and v are the Young's modulus and Poisson's ratio, respectively.

Considering the uniaxial stiffness given by Eq. (A.1), the height of neutral axes, z_x and z_y, are calculated for the flange width of $S_x/(1 - v^2)$ and

$S_y/(1 - v^2)$, respectively. By applying the Kirchhoff–Love assumption, the inplane strains of both deck and bottom plate panels are given by

$$\varepsilon_x = \frac{\partial u}{\partial x} = -(z + z_x)\frac{\partial^2 w}{\partial x^2} \tag{A.2}$$

$$\varepsilon_y = \frac{\partial v}{\partial y} = -(z + z_y)\frac{\partial^2 w}{\partial y^2} \tag{A.3}$$

$$\gamma_{xy} = \frac{\partial u}{\partial y} + \frac{\partial v}{\partial x} = -(2z + z_x + z_y)\frac{\partial^2 w}{\partial x \partial y} \tag{A.4}$$

The axial stresses of stiffeners and webs in both the x- and y-directions are

$$\sigma_{xs} = -E(z + z_x)\frac{\partial^2 w}{\partial x^2} \tag{A.5}$$

$$\sigma_{ys} = -E(z + z_y)\frac{\partial^2 w}{\partial y^2} \tag{A.6}$$

The bending and torsional moments are obtained by integrating the stress components over the cross section as

$$M_x = \int_{A_p} (z + z_x)\sigma_{xp}\,dA_x + \int_{A_{sx}} (z + z_x)\sigma_{xs}\,dA_x \tag{A.7}$$

$$M_y = \int_{A_p} (z + z_y)\sigma_{yp}\,dA_y + \int_{A_{sy}} (z + z_y)\sigma_{ys}\,dA_y \tag{A.8}$$

$$M_{xy} = -\int_{A_x} (z + z_x)\tau_{xy}\,dA_x \tag{A.9}$$

The difference of the height of the neutral axes, z_x and z_y, is usually small. Therefore, they are approximated by an average value as

$$z_x, z_y \cong (z_x + z_y)/2 = z_n \tag{A.10}$$

The integrations of Eqs (A.7), (A.8) and (A.9) based on the assumption of Eq. (A.10) yield the following coefficients in Eq. (4.1):

$$D_x S_x = \frac{E}{1 - v^2}I_{px} + E(I_{sx} + I_{wx}) \tag{A.11}$$

$$D_y S_y = \frac{E}{1 - v^2}I_{py} + E(I_{sy} + I_{wy}) \tag{A.12}$$

$$D_{xy}S_x = \frac{Ev}{1-v^2}I_{px}, \quad D_{yx}S_y = \frac{Ev}{1-v^2}I_{py} \tag{A.13}$$

$$G_{xy}S_x = \frac{E}{2(1+v)}I_{px}, \quad G_{yx}S_y = \frac{E}{2(1+v)}I_{py} \tag{A.14}$$

where I_{px} is the moment of inertia of the deck and bottom plate panels of width S_x with respect to the neural axis $z = z_n$, I_{sx} is the moment of inertia of the deck and bottom stiffeners, A_{sx}, with respect to the neural axis $z = z_n$ and I_{wx} is the moment of inertia of the web plate, A_{wx}, with respect to the neural axis $z = z_n$. I_{py}, I_{sy} and I_{wx} are similar inertia values for the cross section perpendicular to the y axis.

References

Det Norske Veritas (1997) Rules for classification of fixed offshore installations.

Fujikubo, M. and Kaeding, P. (2002) "New simplified approach to collapse analysis of stiffened plates," *Marine Structures*, 15, 251–283.

Fujikubo, M. and Yao, T. (2001) "Structural modeling for global response analysis of VLFS," *Marine Structures*, 14, 295–310.

Fujikubo, M., Nishimoto, M., and Tanabe, J. (2005) "Collapse analysis of very large floating structure in waves using idealized structural unit method," *Proceedings of the 18th Ocean Engineering Symposium*, The Society of Naval Architects of Japan, CD-ROM, Tokyo, Japan.

Fujikubo, M., Xiao, T.Y., and Yamamura, K. (2003a) "Structural safety assessment of a pontoon-type VLFS considering damage to the breakwater," *Journal of Marine Science and Technology*, 7, 119–127.

Fujikubo, M., Olaru, V.D., Yanagihara, D., and Matsuda, I. (2003b) "Collapse analysis of a pontoon-type VLFS in waves," *Proceedings of the 4th International Workshop on Very Large Floating Structures*, Tokyo, Japan, 185–192.

Inoue, K. (2003) "Stress analysis of detailed structures of Mega-Float in irregular waves using entire and local structural models," *Proceedings of the 4th International Workshop on Very Large Floating Structures*, Tokyo, Japan, 219–228.

ISSC (2006) "Report of Committee VI.2 very large floating structures," *Proceedings of the 16th International Ship and Offshore Structures Congress*, Southampton, UK, 2, 391–442.

Kashiwagi, M. (1996) "A B-Spline Galerkin method for computing hydroelastic behavior of a very large floating structure" *Proceedings of the Second International Workshop on Very Large Floating Structures*, Hayama, Japan, 149–156.

Katsura, S., Okada, H., Masaoka, K., and Tsubogo, T. (2003) "A study on structural design of VLFS based on collapse behavior and reliability analysis," *Proceedings of the 4th International Workshop on Very Large Floating Structures*, Tokyo, Japan, 201–208.

Ma, J. and Webster, W.C. (1999) "Optimization of the strength distribution for a model of a large-scale floating runway," *Proceedings of the 3rd International Workshop on Very Large Floating Structures*, Honolulu, Hawaii, USA, 586–593.

Moan, T. (2004) "Safety of floating offshore structures," *Proceedings of the 9th Symposium on Practical Design of Ships and Other Floating Structures*, Luebeck-Travemuende, Germany, 1, 10–37.

Muragishi, O., Yoshikawa, T., Sano, A., Kohsaka, A., and Saoka, S. (1997) "Damage analysis of super large floating structure in airplane collision," *Journal of the Kansai Society of Naval Architects, Japan*, 228, 181–189 (in Japanese).

Ohmatsu, S. (1997) "Numerical calculation of hydroelastic responses of pontoon type VLFS," *Journal of the Society of Naval Architects of Japan*, 182, 329–340 (in Japanese).

Oka, M., Oka, S., Masanobu, S., Kawabe, H., and Inoue, K. (2003) "Wave-induced stress analysis for detailed structural member on very large floating structure (mega-float)," *Proceedings of the 4th International Workshop on Very Large Floating Structures*, Tokyo, Japan, 249–252.

Okada, S., Shibuta, S., Nagayama, H., and Okumura, H. (1999) "Hydroelastic response and structural analysis of a 1000 m Mega-Float model," *Proceedings of the 18th International Conference on Offshore Mechanics and Arctic Engineering*, OMAE99, St. John's, Canada, Paper No. OSU-3060.

Seto, H., Ochi, H., Ohta, M., and Kawakado, S. (2003) "Hydroelastic response analysis of real very large floating structures in regular waves in open/sheltered sea," *Proceedings of the 4th International Workshop on Very Large Floating Structures*, Tokyo, Japan, 65–73.

Suzuki, H. and Yoshida, K. (1996) "Design flow and strategy for safety of very large floating structure," *Proceedings of the Second International Workshop on Very Large Floating Structures*, Hayama, Japan, 21–27.

TRAM (1997) Technological Research Association of Mega-Float "Analysis of strength against airplane collision" (in Japanese).

TRAM (1999) Technological Research Association of Mega-Float "Technical Guide Line of Mega-float" (in Japanese).

Ueda, Y. and Rashed, S.M.H. (1984) "The Idealized Structural Unit Method and its application to deep girder structures," *Computers and Structures*, 18, 277–293.

Yasuzawa, Y. (2003) "Structural optimization of pontoon type VLFS at initial design stage," *Proceedings of the 4th International Workshop on Very Large Floating Structures*, Tokyo, Japan, 209–217.

Yoshida, K. (2005) "Evaluation on and simulation of fires inside large floating structures," *Proceedings of the 18th Ocean Engineering Symposium, Soc of Naval Architects of Japan* (in Japanese), Tokyo, Japan.

Chapter 5

Analysis and design of station-keeping systems

Shigeru Ueda

5.1 Introduction

To date, many floating structures have been constructed for different applications such as floating breakwaters, mooring buoys, wave observation buoys, telemetry buoys, oil drilling platforms, floating piers, floating ferry terminals, offshore oil reservoirs, exhibition halls, floating emergency bases and floating bridges (Shirai 1994; Ueda et al. 2002). Water depths at the construction site may vary from a few meters to more than a few kilometers. When designing the station-keeping system of a floating structure, the actions due to the wind and waves in a stormy weather and earthquakes must be considered. The station-keeping system must be designed to restrict the motions of the floating structure within the allowable values which correspond to the purpose and function of the floating structure as well as to ensure that the structure is safe.

The station-keeping system of a floating structure may be grouped into two main types: (1) the mooring-lines type; and (2) the caisson or pile-type dolphins with fender system. The station-keeping method by mooring lines uses chains, wire ropes, synthetic ropes, chemical fiber ropes, steel pipe piles and hollow pillar links as shown in Figure 5.1. However, the motions of a floating structure become large when the length of mooring line is rather long. Especially in deep seas, the leg system is adopted to which the pretension is applied to the mooring line in order to restrain heaving motion. In such a station-keeping system, it is difficult to restrain the horizontal motion and usually the mooring lines experience significant tension forces.

On the other hand, the dolphin-fenders type is very effective in restraining the horizontal displacement of the floating structure. As the large-size rubber fenders are able to undergo a large deformation (of up to approximately half their lengths), a considerable amount of the kinetic energy of the floating structure can be absorbed. When a pile structure is used for the dolphin, the energy absorption due to the elastic deformation of the pile is ignored because the deformation of the pile is small in comparison with the rubber fenders. In this case, the reaction forces of the rubber fenders become the design

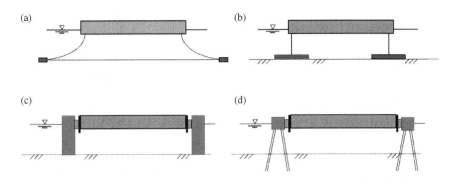

Figure 5.1 Station-keeping systems of floating structure. (a) Chains, ropes and anchors, and sinkers. (b) Tension legs. (c) Caisson dolphin and fenders. (d) Jacket, pile and fenders.

load for the dolphin. Note that in the dolphin-fender mooring system, the motions in the vertical direction are usually free and thus the fender motions only have the other five components of the rigid-body motion.

The dolphin-fender mooring system was first adopted for the two offshore oil reservoirs at Kamigoto (Ikegami and Shuku 1994) and Shirashima Oil Stockpiling Stations (Ito et al. 1994) in Japan. The mooring system has since been used for other facilities such as floating piers, floating terminals, exhibition halls, floating disaster protection facilities and floating bridges. Figure 5.2 shows rubber fenders attached to the caisson-type dolphin of the Shirashima Oil Stockpiling Station. Figure 5.3 shows the rubber fenders attached to the slab of the Yume-Mai Bridge in Osaka City. Figure 5.4 shows a schematic drawing of the jacket-type dolphin and fender system. For a small floating pier, only piles and roller fenders are used as shown in Figure 5.5.

In this chapter, the design and the construction of station-keeping systems with dolphins will be discussed.

5.2   Load-deformation characteristics of devices

5.2.1   Characteristics of rubber fenders

Load-deformation characteristics of rubber fenders

For the dolphin-fender mooring system, the fender characteristics hold the key to the success or failure of the system. Therefore, high-performance rubber fenders have been developed recently. These fenders have been

Figure 5.2 Rubber fenders attached on caisson-type dolphin.
Source: Shirashima Oil Stockpiling Station.

Figure 5.3 Rubber fenders attached on Yume-Mai bridge.

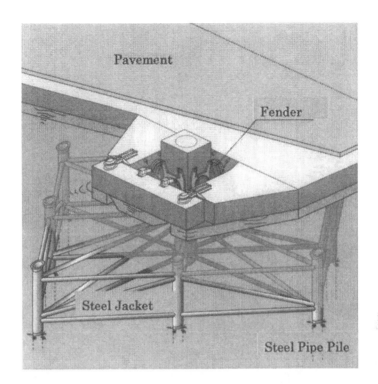

Figure 5.4 Jacket-type dolphin and fender system.

Figure 5.5 Piles and roller fender system.

installed at oil terminals receiving 200,000–500,000 DW crude oil carriers for absorbing ships' berthing energy. High performance fenders, of lengths varying from 3 to 4 m, are capable of withstanding reaction forces between 5,700 and 6,900 kN and the energy absorption amounting to 6,600–7,000 kJ.

The load-deformation characteristics of rubber fenders may be classified under the buckling-type fender and the side-loaded cylindrical-type or air-type fender as shown in Figure 5.6 (PIANC 2002). For the buckling-type

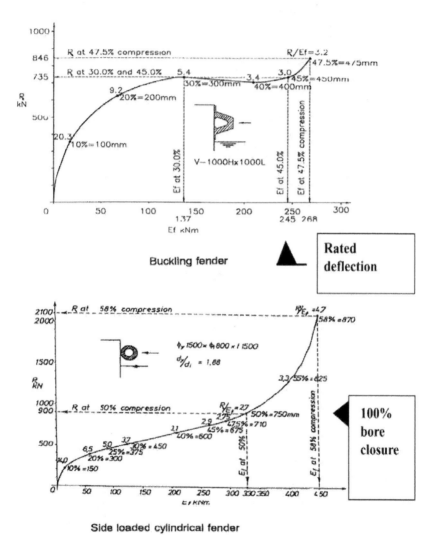

Buckling fender

Side loaded cylindrical fender

Figure 5.6 Effects achieved at various degrees of compression of a side loaded fender.
Source: PIANC 2002.

fender, the reaction force increases rapidly with respect to small deformation and reaches the maximum value at deformation of 20% to 25% of the length. After this point, the reaction force remains almost equal to the maximum reaction force until the deformation reaches 50% to 60% of the length. On the other hand, the reaction force increases exponentially with respect to the deformation for the side-loaded cylindrical-type fender.

The energy absorption E_f (in kJ) by a rubber fender may be defined as (PIANC WG33 2002)

$$E_f = f \times R_m \times d_m \tag{5.1}$$

where f is the factor representing the energy-absorbing efficiency of the fender system (i.e. it varies from 0 to 1), R_m the maximum fender reaction force (in kN) and d_m the maximum fender deformation (in m). The factor f for the buckling-type fender is larger than that of the side-loaded type fender. The factor f is equal to the shaded area divided by the rectangular area $O-R_m-A-d_m$ as shown in Figure 5.7. Therefore, the reaction force of the buckling-type fender is smaller than that of the side-loaded cylindrical-type fender of the same height for the same energy absorption. If the reaction forces of both fender-types are kept the same, then the height of the side-loaded cylindrical-type fender and the allowable deformation may be larger than those of the buckling-type rubber fenders.

The energy absorbed by the fender system during compression is partially returned to the floating structures and partially dissipated in the form of heat within the material (hysteresis in Figure 5.8).

The buckling-type rubber fender is suited for restraining floating structures which are subjected to waves, wind and current which can be modeled as steady forces. Consequently, it is said that the buckling-type fender is suitable for the dolphin-fender-type mooring system.

Table 5.1 shows the variation of the reaction force characteristics of the rubber fender that is used for the station-keeping system of a floating

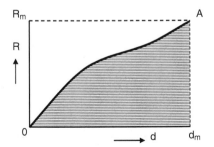

Figure 5.7 Shaded area represents the energy absorption and factor f is equal to the shaded area divided by the rectangular area $O-R_m-A-d_m$.

Source: PIANC (2002).

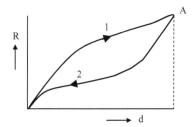

Figure 5.8 Curve 1 represents compression of fender, Curve 2 the decompression of fender, whereas the area between these two curves corresponds to the energy dissipated (warmth generated) as a result of hysteresis.

Source: PIANC (2002).

Table 5.1 Variation of reaction force against the nominal load-deformation characteristics

	Oil Stockpiling Station[a]	Yume-Mai Bridge[b] (2002)
Manufacturing error	0.90–1.10	0.95–1.05
Aging	1.00–1.05	1.00–1.05
Velocity factor	1.00–1.10	1.00–1.05
Creep	The steady load or the mean load shall be less than the reaction force at 10% strain, and to use the load-deflection characteristics in consideration of creep.	
Repeated compression	0.8–0.9 (at 40% deflection repeated for more than 10 cycles)	0.9–1.0 (at 20% deflection repeated for more than 10 cycles)
Inclination compression	Load-deflection characteristics in consideration of the lateral force as 10% of the axial force	0.95–1.00
Temperature factor	0.95–1.25 (at 0–50°C)	0.95–1.25 (at 15–45°C)

Notes
a Japan Port & Harbour Association.
b Bridge Bureau City of Osaka.

structure. The variation described in Table 5.1 was used in the design of the mooring systems for the Kami-Goto and Shirashima Oil Stockpiling Stations (Ito et al. 1994) and Yume-Mai Bridge (Osaka City 2002). The calculations of the motions and mooring forces are made by the numerical simulation method with the fender characteristics variation. The upper and lower limits for the reaction force are determined from Table 5.1. The upper limit of fender characteristics is adopted to estimate the mooring force, while the lower limit of fender characteristics is used to estimate the motions and deflections of the mooring systems. A detailed explanation on major items is given in Table 5.1 (Ueda et al. 1998).

Repeated compression characteristics of the rubber fender

Motions of a floating structure are irregular and periodic due to the actions of waves, wind and current. Therefore, the rubber fender is subjected to a compressive deformation repeatedly. When the material of the structural component is subjected to the repeated-loads action, the examination of its fatigue strength is necessary. Because of this, the reaction force of a fender which is used for the station-keeping systems is lowered to 0.8–0.9 times that of the normal reaction force when the fender is repeatedly compressed at 40% deformation of fender height for more than 10 cycles. This provision is prescribed based on the result of a repetition compression test which is shown in Figure 5.9. The reaction force drops around 10% when a fender is compressed 1000 cycles repeatedly at the 20–30% deformation and approximately 10–20% when it is compressed 10–100 cycles at the 50% deformation. In Figure 5.9, the symbol RH means a rubber kind.

Temperature factor

The temperature factor for a certain kind of rubber is illustrated in Figure 5.10.

The temperature factor varies for different kinds of rubber. As the rubber becomes remarkably stiff at low temperatures, it is necessary to determine the load-deformation characteristics of the rubber fender under the

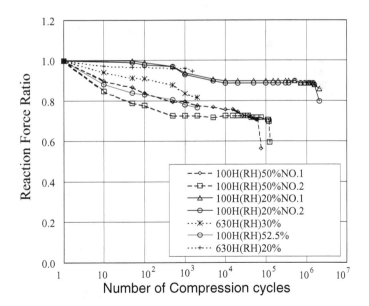

Figure 5.9 Repeated compression characteristics of rubber fender.

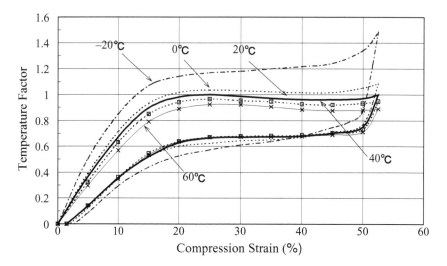

Figure 5.10 Temperature factor for SUC100H(RH).

local temperature condition before performing numerical simulations for determining the motions and mooring forces of a floating structure. In Figure 5.10, SUC denotes a fender type and 100 H represents the size of the model fender.

Velocity factor

The velocity factor of a particular kind of rubber is illustrated in the dynamic load-deformation characteristics as shown in Figure 5.11. The velocity factors vary not only with the type of rubber but also with the temperature factor.

The compression test of a large sized rubber fender is usually done at a load speed of 6–8 cm/min according to the capacity of the existing testing machine. However, fenders in the field can be compressed at a faster speed than that used in the compression test. Therefore, it is necessary to examine the velocity factor by experiments. The velocity factor must be incorporated in the numerical simulations for determining the motions and mooring forces of a floating structure.

The variation of the velocity factor is in the range of 1.0–1.1. This range is adopted in the design of the Offshore Oil Stockpiling Stations. However, attention is needed when using these velocity factors as they vary for different kinds of rubber.

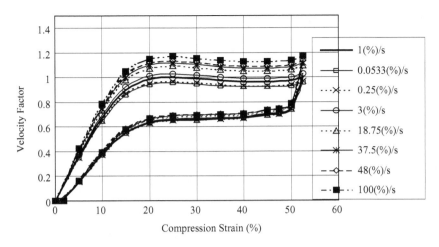

Figure 5.11 Velocity factor for SUC100H(RH).

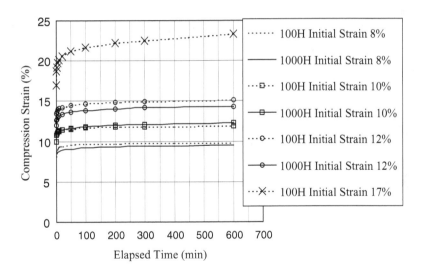

Figure 5.12 Static creep for SUC100H(RH).

Creep characteristics

Figure 5.12 shows that the compression strainchange with respect to time for a buckling-type rubber fender when subjected to a steady load. Table 5.2 is the summary of the creep tests. The load is set as the reaction force against those deformations equivalent to 8%, 10%, 12%, and 17% strain. It is understood that it becomes too unstable when the steady load or the mean

Table 5.2 Summary of static creep after 10 h

Initial strain (%)	SUC1000H (%)	SUC100H (%)
8	9.50	9.80
10	12.30	11.90
12	14.30	15.10
17	—	23.30

Source: Ueda et al. (1998).

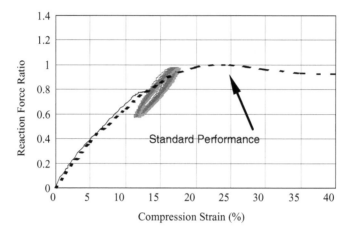

Figure 5.13 Image of dynamic creep.

load act on the rubber fender is larger than the reaction force against 10% strain. Therefore, it is advisable to select a rubber fender such that the steady load or the mean load is less than the reaction force against 10% strain. This point should be incorporated in the numerical simulations for calculating the motions and mooring forces of a floating structure when using the load-deformation characteristics in allowance for the effect of creep. On the other hand, dynamic creep will be developed for the repeat compression. Figure 5.13 shows the image of the dynamic creep. The dynamic creep may be less than 2% when the initial deformation is less than 10% strain.

Other factors

Other than the aforementioned factors, it is also necessary to incorporate other characteristic variations on the rubber fender, such as the manufacturing error, aging and the inclination compression-character.

5.2.2 Load-deformation characteristics of mooring dolphin

The structural type of mooring dolphins is broadly classified under the gravity-type structure, such as a caisson and a cellular bulkhead, and the pile-type structure, such as a vertical-pile pier, a coupled-pile pier and a jacket type. A gravity-type dolphin is dealt with as a rigid body and is designed so that the interaction force between the dolphin and the mooring systems does not exceed the resistance force for sliding. A pile-type structure appears to be an elastic body but it is treated as a rigid body because its rigidity is much larger than that of the rubber fender.

A high-tensile steel is often used for piles of flexible dolphins in order to make use of the energy absorption by the dolphin itself. The complex load-deformation characteristics of both piles and fenders should be adopted in the numerical simulations for determining the motions and mooring forces of a floating structure.

The load-deformation characteristics in the horizontal direction of the pile-type dolphin may be calculated by various methods the method by Blum, or by Chang (1937), or by Matlock (1970) and Reese et al. (1975), which conforms with the API method (1976) and by Kubo (1964), and Hayashi and Miyajima (1963), which conforms tothe PHRI method (1996); by FEM, and so on.

5.3 Estimation of mooring force

5.3.1 Outline

The mooring force of a floating structure subjected to waves or seismic motion is usually determined by means of a numerical simulation. The equation of motions of a moored floating structure takes on a second-order differential equation with six degrees of freedom. The time-domain numerical simulation is done in consideration of forces from irregular waves, the second order wave force or so-called steady or fluctuating drift force, current force, fluctuating wind force, seismic force and the nonlinear load-deformation characteristics including hysteresis of mooring facilities (see Figure 5.3).

After the statistical analysis of the time-domain numerical simulation results, quantities such as the maximum value, the maximum amplitude, the significant value and the mean of motions, mooring forces, the response acceleration and velocity are calculated.

5.3.2 Numerical simulation method

The numerical simulation of a floating structure is a time-domain analysis of second-order differential equation described by the following equation involving six degrees of freedom, namely surge, sway, heave, roll, pitch and yaw:

$$\sum_{j=1}^{6}(M_{ij} + m_{ij}(\infty))\ddot{X}_j(t) + \sum_{j=1}^{6}\left\{\int_{-\infty}^{t} L_{ij}(t - \tau)\dot{X}_j(\tau)d\tau + D_i(t)\right\}$$

$$+ \sum_{j=1}^{6}(C_{ij} + G_{ij})X_j(t) = F_i(t) \tag{5.2}$$

where M_{ij} is the mass of floating structure; $m_{ij}(\infty)$ the added mass; $X_j(t)$ the motion of floating structure at time t; $L_{ij}(t)$ the retardation function at time t; $D_i(t)$ the damping force due to mooring lines and viscosity at time t; C_{ij} the restoring force coefficient; G_{ij} the mooring force coefficient; $F_i(t)$ the external forces at time t; i, j the mode of motions of floating structure ($i, j = 1$ to 6).

The retardation function and the added mass are calculated by using the following relations:

$$L_{ij}(t) = \frac{2}{\sigma} \int_{0}^{\infty} B_{ij}(\sigma) \cos \sigma t \, d\sigma \tag{5.3}$$

$$m_{ij}(\infty) = A_{ij}(\sigma) + \frac{1}{\sigma} \int_{0}^{\infty} L_{ij}(t) \sin \sigma t \, dt \tag{5.4}$$

where σ is the angular frequency; $A_{ij}(\sigma)$ the added mass at angular frequency σ; and $B_{ij}(\sigma)$ the damping coefficient at angular frequency σ.

The motions of the floating structure are determined by the predominant angular frequency of external forces. At the natural-angular frequency, the frequency characteristics are related to the natural period of the moored floating structure. Therefore, the radiation forces are usually represented by the value at a significant wave period or at a natural period of motion.

When the spectrum of external forces is not wide band, motions of floating structure are calculated by the following equation:

$$\sum_{j=1}^{6}(M_{ij} + A_{ij}(\sigma_0))\ddot{X}_j(t) + \sum_{j=1}^{6}\left\{\int_{-\infty}^{t} B_{ij}(\sigma_0)\dot{X}_j(\tau)d\tau + D_i(t)\right\}$$

$$+ \sum_{j=1}^{6}(C_{ij} + G_{ij})X_j(t) = F_i(t) \tag{5.5}$$

where the added mass and the damping coefficient for radiation forces are calculated at the angular frequency σ_0 such as the predominant frequency of external forces or the natural frequency.

5.3.3   Loads and forces act on a floating structure

Loads and external forces acting on a floating structure are the selfweight, buoyancy and external forces such as wave force, wind force, current force, seismic force, and so on. With the action of those loads and forces, motions of a floating structure are developed, the mooring system deformed and reaction forces generated. The load action is given on the right-hand side of Eq. (5.5). The motions of a floating structure and mooring forces of the station-keeping system are calculated by numerical simulations.

As loads such as wind force, wave force and seismic force are irregular and periodic, the frequency characteristics must be considered in the numerical simulation. Below, is an outline of the treatment of wind force, wave force and seismic force.

Wind force

Generally wind speed is given as the average wind speed. A wind speed varies with respect to time and space and the maximum instantaneous wind speed is usually larger than the average wind speed. The ratio of the maximum instantaneous wind speed and the average wind speed of a certain point is called the gust ratio.

Though it is appropriate that the frequency spectrum of wind shall be determined according to the observed data in a construction site, however, if there is no observed data, the frequency spectrum of wind speed by Davenport (1967), Hino (1976) or any other type can be used in the numerical simulations as an irregular and periodic wind speed. Equations (5.6) and (5.7) are the frequency spectrum of wind speed by Davenport and Hino, respectively,

$$fS_u(f) = 4K_r U_{10}^2 \frac{X^2}{(1+X^2)^{4/3}}, \quad X = \frac{1200f}{U_{10}} \tag{5.6}$$

$$S_u(f) = 2.856 \frac{K_r U_{10}}{\beta} \left\{ 1 + \left(\frac{f}{\beta} \right)^2 \right\}^{-5/6}$$

$$\beta = 1.169 \times 10^{-3} \left(\frac{U_{10}\alpha}{\sqrt{K_r}} \right) \left(\frac{z}{10} \right)^{2m\alpha-1} \tag{5.7}$$

Wind forces that are acting on a floating structure can be calculated by using the following equations:

$$R_X = \tfrac{1}{2} \rho_a U^2 A_T C_X \tag{5.8}$$

$$R_Y = \tfrac{1}{2} \rho_a U^2 A_L C_Y \tag{5.9}$$

$$R_M = \tfrac{1}{2} \rho_a U^2 A_L L C_M \tag{5.10}$$

where C_X is the drag coefficient in the X direction (i.e. from the front of a floating structure); C_Y the drag coefficient in the Y direction (i.e. from the side of a floating structure); C_M the pressure-moment coefficient about the center of gravity of a floating structure; R_X the component of resultant wind force (kN) in the X direction; R_Y the component of resultant wind force (kN) in the Y direction R_M the moment (kJ) of resultant wind force about the center of gravity of a floating structure; ρ_a the density of air $= 1.23 \times 10^{-3}$ (t/m^3); U the wind speed (m/s); A_T the front-projected area above the water surface (m^2); A_L the side-projected area above the water surface (m^2); and L the length of a floating structure (m). Note that the coefficients C_X, C_Y, and C_M are determined by a wind tunnel test.

Wave force

The wave force is a force exerted by the incident waves on a floating structure when the floating structure is considered to be fixed in the water. It is composed of a linear force that is proportional to the amplitude of the incident waves and a nonlinear force that is proportional to the square of the amplitude of the incident waves. The linear force is the force that the floating structure receives from the incident waves as the reaction when it deforms the incident waves. It is expressed as the sum of the Froude–Krylov force and the diffracted wave force. As sea waves are irregular and periodic, the wave force is given as a time series of irregular and periodic value. For the numerical simulations, one may use the frequency spectrum by Bretshneider (1968) and Mituyasu (1970) given by:

$$S(f) = 0.257 H_{1/3}{}^2 T_{1/3} (T_{1/3} f)^{-5} \exp\left[-1.03(T_{1/3} f)^{-4}\right] \tag{5.11}$$

Where $S(f)$ is the frequency spectrum; f the frequency; $H_{1/3}$ the significant wave height and $T_{1/3}$ is the significant wave period. The equation was originally proposed by Bretshneider and was modified by Mituyasu on the condition that there was the relation of $T_p \cong 1.0 T_{1/3}$ between the peak period and significant wave period. Goda (1987) proposed Eq. (5.12) in correcting the relation between the peak period and significant wave period as $T_p \cong 1.1 T_{1/3}$ according to observed data.

$$S(f) = 0.205 H_{1/3}{}^2 T_{1/3}^{-4} f^{-5} \exp\left[-0.75(T_{1/3} f)^{-4}\right] \tag{5.12}$$

Those wave spectrums of Eqs (5.11) and (5.12) may be applied when the wind field is stationary. However, JONSWAP spectrum is applied to wind waves which grow in rather short fetch-length under strong winds. Other types of frequency spectrum may also be used.

Radiation forces are induced accordingly to develop motions of a floating structure. The radiation force is divided into two components that are proportional to the acceleration and velocity. They are, respectively, treated as added mass and damping coefficient in the equation of motions.

The wave-drift force which is proportional to the square of the wave height must be considered when the length of a floating structure becomes equal or larger than the wavelength. By assuming that the floating structure is two-dimensional and the wave energy is not dissipated, the wave-drift force is given by:

$$F_d = \frac{1}{8}\rho_w g H_i^2 R$$

$$R = K_R{}^2 \left\{ 1 + \frac{4\pi h/L}{\sinh\left(4\pi h/L\right)} \right\}$$

(5.13)

where F_d is the wave drift force per unit length (kN); ρ_w is the density of the seawater (kg/m^3); H_i is the wave height of incident wave (m); K_R is the ratio of reflection and R is the coefficient of wave drift force.

Seismic load

When a floating structure is located in a seismic region, the station keeping system has to be designed with consideration of seismic forces. In such places, the floating structure is subjected to seismic forces which are proportional to its acceleration and mass as well as the interaction force between itself and the station-keeping system.

The relative response acceleration of the floating structure is small because the natural period of a moored floating structure may be larger than the predominant period of the seismic motion. However, since the relative displacement with the mooring system is large, it is recommended that mooring systems with appropriate stiffness be selected with the condition of use in mind.

5.3.4 Numerical simulation and statistical analysis of results

Motions and mooring forces calculated by means of numerical simulations and results of the hydraulic experiment areofirregular and periodic value. Yet, the expected value of the variables is to be estimated in the duration of rough weather condition because generally the elapsed time of the numerical simulation or the hydraulic experiment is limited. Results from the computational analysis and the hydraulic experiment shall be analyzed and presented in the form of time histories with the maximum values, the significant values, the average values, the frequency spectrum, the expected values, and so on.

In consideration of both frequency characteristics of external forces and the response characteristics of the moored floating structure, it is recommended that the time step of the numerical simulation be set at about 1/8 of the minimum period of the external forces or less. It is also recommended that the numerical simulation be performed as long as possible when more than one hundred effective amplitudes of motions and mooring forces are to be obtained in order to calculate the expected values accurately.

5.4 Design of station-keeping systems

5.4.1 Design of gravity-type dolphin

The stability of the gravity-type dolphin shall be examined with respect to sliding, overturning, and bearing capacity of the foundation and settlement. The safety factor F_S against sliding of a gravity type dolphin may be calculated using the following equation:

$$F_S \le \frac{fW}{P} \tag{5.14}$$

where W is the resultant vertical force acting on the dolphin (kN); P the resultant horizontal force acting on the dolphin (kN); and f the coefficient of friction between the bottom of the dolphin and the foundation. In this case, the resultant horizontal resistance is given as the product of the resultant vertical force and the coefficient of friction between the bottom of the dolphin and the foundation. The resultant vertical force is calculated by considering the surcharge, selfweight of the dolphin, buoyancy, seismic force, dynamic water pressure during earthquakes and interaction force (i.e. mooring force from the station-keeping systems).

On the other hand, the safety factor F_S against the overturning of the gravity-type dolphin may be calculated by the following equation:

$$F_S \le \frac{Wt}{Ph} \tag{5.15}$$

where t is the distance between the line of application of the resultant vertical forces acting on the dolphin and the front toe of the dolphin (m); h the height of the application of the resultant vertical forces acting on the dolphin, above the bottom of the dolphin (m). Note that the safety factor in Eq. (5.14) is given as the ratio of the resultant resistant moment and the resultant overturning moment.

The bearing capacity is assessed based on the foundation materials, the size of structures, and so on. The circular arc analysis is used when examining the bearing capacity for eccentric and inclined loads acting on a foundation of gravity-type structures. Details on the formulae of bearing capacities are available in the codes or the design recommendations for port and harbor facilities and offshore structures.

5.4.2 Design of pile-type dolphin

The design of the pile-type dolphin involves the examination of the axial bearing capacity and the lateral bearing of piles as well as the determination of the pile dimensions.

The ultimate-axial bearing capacity of pile is given by:

$$Q_d = Q_f + Q_p = fA_s + qA_p \tag{5.16}$$

where Q_d is the ultimate axial-bearing capacity of pile (kN); Q_f the bearing capacity by circumferential skin-friction (kN); Q_p the toe-bearing capacity (kN), f the mean circumferential skin-friction intensity (kN/m^2); A_s the total circumferential area of pile (m^2); q the toe-bearing capacity intensity (kN/m^2) and A_p the toe area of pile (m^2). The appropriate values for f and q are given by the codes or recommendations by authorized institutions and societies.

The basic equation for determining the behavior of a lateral pile modeled as a beam on an elastic foundation is given by:

$$EI\frac{d^4y}{dx^4} + Bp(x,y) = 0 \tag{5.17}$$

where $p(x,y)$ is the subgrade reaction per unit area at depth x and with displacement y (kN/m^2); B is the pile width (m); EI the flexural rigidity of pile (kNm^2); x is the depth from the ground (m); and y is the displacement of pile at the depth (m).

There are several methods for evaluating the subgrade reaction force. These methods include the earth pressure theory under the ultimate equilibrium condition of the soil and the elastic subgrade methods proposed by Chang (1937), Kubo (1964), and Hayashi and Miyajima (The Japan Port & Harbour Association 1999b) which conforms to the PHRI method, and by Matlock and Reese which conforms to the API method.

The subgrade reaction per unit area at depth x and with displacement y is given by:

$$p(x,y) = k_h x^m y^n \tag{5.18}$$

where k_h is the factor of lateral resistance of ground. It is so called linear elastic-subgrade method or nonlinear subgrade method when $n = 1$ or $n \neq 1$ respectively while index m, n are $m \geq 0$ and $0 \leq n \leq 1$. The solution is given by Chang (1937) when $m = 0$ for the elastic-subgrade method, and it is well known because of its simplicity and easy treatment. However, it is worth noting that the subgrade reaction will be overestimated when compared with the results of other methods when the lateral displacement of the pile is large. On the other hand, Kubo defined the values of index m, n as $m = 1$, $n = 0.5$ for sandy soil while Hayashi and Miyajima defined these values as $m = 0$, $n = 0.5$ for clay-like soil according to experimental results.

As for the method by Chang (1937), the following equations are to be solved:

above the ground surface $EI\dfrac{d^4y_1}{dx^4} = 0$ $(0 \geq x \geq -h)$ (5.19)

under the ground surface $EI\dfrac{d^4y_2}{dx^4} + E_s y_2 = 0$ $(x \geq 0)$ (5.20)

where $E_s (=k_h B)$ is the modulus for the ground (kN/m^2); B is the pile width (m); EI the flexural rigidity of pile (kNm^2); x is the depth from the ground (m); and y_1, y_2 are the displacement of pile at the depth of x (m). Note that Chang's method is based on two assumptions: (1) the factor of lateral resistance of ground k_h is constant at any depth of y; and (2) the length of pile is long enough.

The boundary conditions are given by

at $x = -h$: $\dfrac{dy_1}{dx} = 0,$ $S = -\dfrac{d^3y_1}{dx^3} = -H$

at $x = 0$: $y_1 = y_2,$ $\dfrac{dy_1}{dx} = \dfrac{dy_2}{dx},$ $\dfrac{d^2y_1}{dx^2} = \dfrac{d^2y_2}{dx^2},$ $\dfrac{d^3y_1}{dx^3} = \dfrac{d^3y_2}{dx^3}$

at $x = \infty$: $y_2 = 0,$ $\dfrac{dy_2}{dx} = 0$

Solutions of Eqs (5.19) and (5.20) are given as follows:

$$y_1 = \frac{H}{12EI\beta^3} \left\{ 2\beta^2 x^3 - 3(1 - \beta h)\beta^2 x^2 - 6\beta^2 hx + 3(1 + \beta h) \right\}$$ (5.21)

$$y_2 = \frac{H}{4EI\beta^3} e^{-\beta h} \left\{ (1 + \beta h) \cos \beta h + (1 - \beta h) \sin \beta x \right\}$$ (5.22)

where the so-called characteristic value

$$\beta = \sqrt{\frac{Bk_h}{4EI}}$$ (5.23)

The value of $1/\beta$ is referred to as the virtual fixed point of pile. According to Yokoyama (1977), piles of finite length can also be calculated in a similar way to piles of infinite length as long as $\beta L \geq \pi$.

By differentiating Eqs (5.21) and (5.22), the deflection angle θ_x the bending moment M_x the shear force S_x and the subgrade reaction force can be obtained.

5.5   Construction of station-keeping systems

5.5.1   Installation of gravity-type dolphin

Mostlya concrete caisson is mostly used for the main structure of the gravity type dolphin. The caisson is usually built at a caisson yard. The caisson yard is classified as having one or more of the following: slipway, dry dock, floating dock, floating-crane lift on and off facilities, an excavated dock, being situated on the sea, and so on according to the method of fabrication and launching.

After the construction of a caisson in the caisson yard, it will be towed or carried to the site for installation. When it is lifted and carried to the installation site, a steel liftframe is used to equally distribute the caisson weight to the lifting steel bars.

The caisson is towed by tugboats or pushed by pusher barges to the installation site in a floating condition. Wire ropes are set surrounding the caisson and/or at the towing metal fittings embedded in the caisson wall. After the caisson is towed and/or carried at the installation site, water is poured into the caisson to sink it. The caisson will be pulled by means of wire ropes which are set between the caisson and the major barge through small barges.

After grounding the caisson, filling is done immediately. Upon completion of the filling process, cover concrete plates and the superstructure are installed.

5.5.2   Installation of pile-type dolphin

The pile-type dolphin—which includes vertical-pile structure, coupled-pile structures, template and jacket-type structures—are installed by driving piles at sea. Piles are driven by means of pile-driving barges, a self-elevating platform (SEP), pile-driving frame with hammer installed on the derrick barge, an offshore hammer set on the top of a pile and an underwater hammer. There are diesel hammers, single steam hammers, hydraulic drop hammers, and multi air gun dumpers.

A template is a kind of steel frame which is held above sea water by derrick barges. A pile is inserted and driven by offshore hammers through the template held by the SEP, derrick barges or floating cranes.

As for the jacket-type structures at water depths around 70–80 m, piles are driven through steel pipe legs of the jacket. However, for a jacket structure installed at a depth deeper than the aforementioned depths, the skirt piles at the foot of the jacket legs are driven by underwater hammers through the sleeve piles. Cement mortar is grouted into the gap between the jacket leg and sleeve pile in order to connect them.

When water depth is shallow, a jacket and a template are not recommended. Instead, piles are driven separately using pile-driving barges or a

SEP which mounts the pile-driving frame. After the piles are driven, the superstructure is installed.

5.6   Maintenance of station-keeping systems

5.6.1   Inspection item

The inspection items for maintenance of the station-keeping systems should be established. A general inspection must be done at least once a year as well as for an extra-ordinary condition such as when strong external forces (due to typhoons as example) act on the facility. The checklists in the form of daily, weekly, monthly, and yearly checks preferably should be prepared. An inspection record must be kept with the description of the treatment details.

5.6.2   Mooring dolphin

For the mooring dolphins, visual inspection must be carried out and the existence of damage checked. The movement and the settlement of the dolphin are measured if present. The corrosion is measured in the case of steel structures.

5.6.3   Rubber fenders

For rubber fenders, existence of damage is checked by visual inspection. Their compression performance is examined if necessary. Conditions of the protector board, installation metal fittings and the base seat are also checked. It is recommended that the compression performance of selected rubber fenders be tested about once n every few years because of deterioration due to aging and the decline in performance due to repetitive loadings.

References

American Petroleum Institute (API) (1976) *API Recommended Practice for Planning, Designing and Constructing Fixed Offshore Platforms*, API RP 2A 7th edition, pp. 21–26.

Bretshneider, C.L. (1968) "Significant waves and wave spectrum," *Ocean Industry*, Feb. 1968, pp. 40–46.

Bridge Bureau, City of Osaka (2002) "Record of construction of Yume-Mai bridge," *City of Osaka*, Appendix p. 45.

Chang, Y.L. (1937) "Discussion on 'Lateral pile loading' by Feagin," *Trans. ASCE*, Paper No. 1959, pp. 272–264.

Davenport, A.G. (1967) "Gust loading factors," *Proc. of ASCE*, ET3, pp. 11–34.

Goda, Y. (1987) "Statistical variability of sea state parameters as a function of wave spectrum," *Coastal Engineering in Japan, JSCE*, Vol. 34, pp. 39–52.

Hino, M. (1976) "Relationship between the instantaneous peak values and the evaluation time – A theory on the gust factor," *Proc. of JSCE*, Vol. 177, pp. 23–33.

Ikegami, K. and Shuku, M. (1994) "Design and field measurement of mooring system for the world's first floating type oil storage system in Kami-Gotoh, Japan," *Proc. of the Floating Structures in Coastal Zone, Port and Harbour Research Institute*, pp. 324–335.

Ito T., H. Chiba and E. Kato (1994) "Main offshore structures of Shirashima floating oil storage terminal," *Proc. Floating Structures in Coastal Zone, Port and Harbour Research Institute*, pp. 103–120.

Japan Port & Harbour Association (1999a) "Technical Standards and Commentaries for Port and Harbour Facilities," *The Japan Port & Harbour Association*, Vol. 1, pp. 455–473.

Japan Port & Harbour Association (1999b) "Technical Standards and Commentaries for Port and Harbour Facilities," *The Japan Port & Harbour Association*, Vol. 2, pp. 1083–1084.

Kubo K. (1964) "A new method for the estimation of lateral resistance of piles," *Report of PHRI*, Vol. 2, No. 3, pp. 1–37.

Matlock, H. (1970) "Correlations for design of laterally loaded piles in soft clay," *Offshore Technology Conference*, Paper No. OTC1204.

Mituyasu, H. (1970) "Wind wave generation and their spectrum-wind wave spectrum in finite fetch length," *Coastal Engineering in Japan*, JSCE, Vol. 17, pp. 39–52.

Permanent International Association Navigation Congress (2002): "PIANC Mar-Com Working Group 33 Guidelines for the Design of Fenders Systems:2002," *Report of WG33, Permanent International Association Navigation Congress*, pp. 1–70.

Reese, L.C., Cox, W.R., and Koop, F.D. (1975) "Analysis of laterally loaded piles in sand," *Offshore Technology Conference*, Paper No. OTC2080.

Shirai, S. (1994) "Introduction of floating facilities in the coastal zone of Japan," *Proc. Floating Structures in Coastal Zone*, Hiroshima, *Port and Harbour Research Institute*, pp. 86–102.

Ueda, S., S. Miyai and N. Masui (2002) "Study on floating bridge in Japan," *Proc. Floating Structures in Coastal Zone, Port and Harbour Research Institute*, pp. 74–85.

Ueda, S., S. Shiraishi, T. Maruyama, A. Uesono, M. Takasaki and S. Yamase (1998) "Properties of rubber fender in use of mooring for floating bridge," *14th Ocean Engineering Symposium, The society of Naval Architect of Japan*, pp. 359–364.

Yokoyama, Y. (1977) "Calculation methods and examples for pile structures," *Sankaido Publishing Co. Ltd.*, p. 69.

Chapter 6

Analysis and design of breakwaters

Tetsuya Hiraishi

6.1 Introduction

Very large floating structure (VLFS) may sometimes be constructed in a location where the sea state is rather harsh such as along the Pacific coastline of Japan. Under such severe wave conditions, the VLFS cannot be moored safely and thus an offshore breakwater is required to reduce the wave forces impacting on the floating structure.

The design wave height and period are estimated from forecasted and observed wave data in rough weather. Several numerical wave-forecasting models are available for this purpose (e.g. The WAMDI Group 1988). For offshore wave observation, a network called NOWPHAS (Nagai and Nukata 2004) is employed for Japanese harbors. The design wave condition is categorized using the return period T_r which indicates the occurrence probability of a target wave height. If the information of the wave period is not sufficient in determining the wave-period probability, the wave period may be derived using the relationship $H/L = 0.04$, where H is the wave height and L is the wave length. Figure 6.1 shows examples of the arrangement of breakwaters for a VLFS moored in a water area protected by breakwaters.

Takada et al. (2002) proposed a table showing the design wave heights along the Japanese coastlines. In the paper, the design wave height and period are presented for different return periods, design tsunami heights and storm-surge deviations in each harbor area. Figure 6.2 shows an example of the design wave height distribution. Each representative offshore point is plotted in the area with a deep-water depth. Usually a breakwater is constructed in shallow water areas and thus, the offshore wave height H_o has to be replaced by the coastal design wave height H. The coastal design wave height H is calculated from the following relation:

$$H = K_r K_d K_s H_o \tag{6.1}$$

where K_r is the wave reflection coefficient, K_d the wave diffraction coefficient and K_s the wave shoaling coefficient. Meanwhile, the equivalent

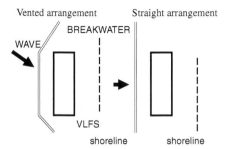

Figure 6.1 Example of arrangement of a VLFS and breakwater.

deep-water wave height H'_o which includes the influence of wave reflection and diffraction in the shallow water is given by

$$H'_o = K_r K_d H_o \tag{6.2}$$

In the design of breakwaters, H'_o is employed as the offshore wave condition. Takada et al. (2002) have also incorporated the influence of wave refraction and diffraction as the coastal coefficient. The coastal coefficient in Figure 6.2 is calculated as the value at the point with the depth of 10 m as derived from the wave-energy balance method.

The return period of the design wave is determined based on the importance and usage condition of the sheltered area. Usually the industrial harbor is designed for a 50-year return period. Therefore, the 50-year return period is suitable for the design of a VLFS. However, a higher return period is recommended when the value of a VLFS is very high, such as a residential complex constructed on the VLFS.

In this chapter, the effectiveness of breakwaters in reducing the wave height acting on a moored VLFS will be discussed. The different types of breakwater including the slope and vertical type are introduced in Section 6.2. In Section 6.3, the wave-action formula for estimating wave force is presented. In Section 6.4, the experimental formula for estimating the wave-transmission coefficient at the breakwater line is demonstrated where it is decreasing behind the breakwater as the height of breakwater crown increases. Section 6.5 describes an experiment for the motion of the VLFS moored behind a breakwater by considering directional waves. It is shown that the construction of the breakwater is effective in reducing the motion of the VLFS. The influence of the overtopping wave is negligible as the relative crown height h_c/H_{in} becomes greater than 0.6.

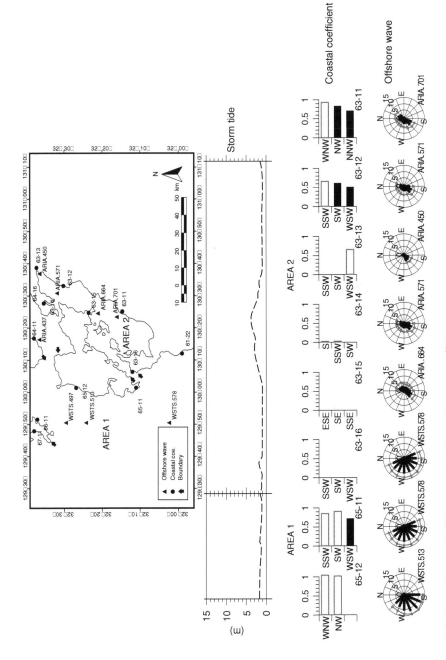

Figure 6.2 Example map of offshore wave heights and coastal coefficients.

6.2   Types of breakwaters

The types of breakwaters sheltering the water area where a VLFS is moored may be grouped under the sloping type and the vertical type. Special types like the curtain-wall breakwater have been proposed in relatively shallower areas. VLFS have been constructed in a relatively deep-water site with a depth more than 10 m. In such a deep offshore area, the sloping- and vertical-type breakwaters are widely employed. Takahashi (1996) proposed the typical patterns of sloping and vertical breakwaters as shown in Figure 6.3(1) and (2). According his criteria, each type has been developed for several reasons.

6.2.1   Sloping type (mound type)

The sloping- or mound-type breakwaters basically consist of rubble mounds as shown in Figure 6.3(1). The most fundamental sloping-type breakwater

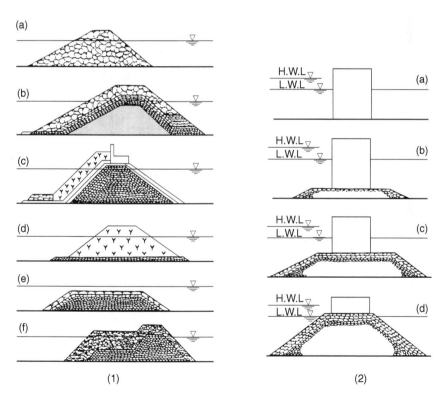

Figure 6.3 Type of breakwater and example of wave-dissipating blocks. (1) Sloping-type breakwaters; (2) vertical-type breakwaters; (3) horizontally composite breakwaters; (4) composite breakwaters and (5) example of wave-energy-dissipating blocks.

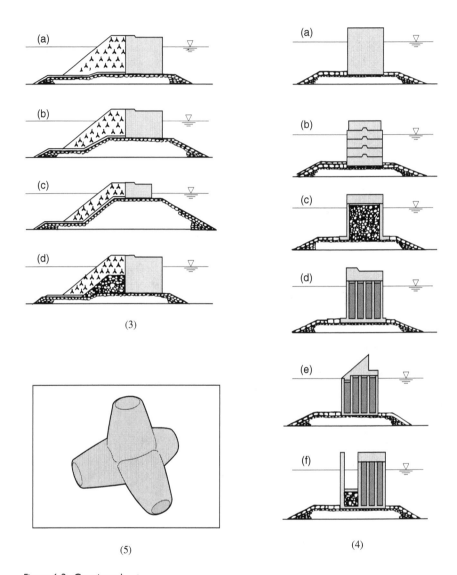

Figure 6.3 Continued.

is one with randomly placed stones as shown in Figure 6.3(1a). In order to increase stability and to decrease wave transmission, the multilayered rubble-mound breakwater was developed with a core made of the fine materials as shown in Figure 6.3(1b). The stability of the armor layer can be strengthened using concrete blocks, while the wave transmission can be reduced using a superstructure (wave screen or wave wall) that can also be a function as an access road to the breakwater as shown in Figure 6.3(1c).

Breakwaters composed of only concrete blocks, as shown in Figure 6.3(1d), are also being constructed, especially as a detached breakwater for coastal protection. Although the wave transmission is not significantly reduced for this breakwater type, its simple construction procedure and the relatively high permeability of the breakwater body are preferable to the water exchange in coastal zone. Recently reef breakwaters or submerged breakwaters, as shown in Figure 6.3(1e), have been constructed for coastal protection as well as to provide the beautiful seascape. The reshaping breakwater, as shown in Figure 6.3(1f), is employed to complete the low transmission and high stability of rubbles. The type is also called a 'berm breakwater.'

6.2.2 Vertical-type (composite and horizontally composite types)

The original concept of the vertical breakwater was to reflect waves, while that for the rubble-mound breakwater was to break them. Figure 6.3(2) shows four vertical-type breakwaters having different mound heights. The basic vertical-wall breakwater is shown in (2a). The others are composite breakwaters with a rubble-mound foundation, namely, the low-mound (2b) and high-mound composite breakwaters (2c). Conventionally, the high-mound composite breakwater has a mound that is higher than the low water level (LWL). The former breakwater does not cause wave breaking on the mound, while the latter one does. Since the high-mound composite type is unstable due to wave-generated impulsive pressure and scouring caused by breaking waves, composite breakwaters with a low-mound are more common. The composite breakwater with a relatively high mound (2c) which is lower than LWL occasionally generates impulsive wave pressure due to wave breaking. The breakwater (2d) has the wave-barrier effect similar to the sloping-type breakwaters.

In order to reduce the wave reflection and the breaking wave force on the vertical wall, concrete blocks are placed in front of it. This composite breakwater covered with wave-dissipating concrete blocks is called the horizontally composite breakwater. Such breakwaters are not new since vertical-wall breakwaters suffering damage to the vertical walls were often strengthened by placing large stones or concrete blocks in front of them. The stones and concrete blocks dissipate the wave energy and reduce the wave force, especially that from breaking waves. Modern horizontally composite breakwaters employ shape-designed concrete blocks such as tetrapods. Figure 6.3(5) shows an example of wave-dissipating blocks.

A horizontally composite breakwater is very similar to a rubble-mound breakwater armored with concrete blocks. Figure 6.3(3) shows the variation of its cross section with respect to the mound height. As the mound height increases, the breakwater becomes very similar to the rubble-mound breakwaters. In particular, a breakwater with core stones in front of the vertical wall (3d) is nearly the same as the rubble-mound breakwater. They are, however, different since the concrete blocks of the rubble-mound breakwater act as an armor for the rubble foundation, while the concrete blocks of the horizontally composite breakwater is set to reduce the wave force and height of the reflected waves.

Thus horizontally composite breakwaters are considered to be an improved version of the vertical types. When a breakwater is designed to protect a VLFS mooring area, the water depth becomes relatively large. This is because, in deeper sea, the VLFS is more effective than the reclaimed-island. The slope-type breakwaters need a wider foot-mound and thus the water area behind breakwater becomes small. In rough weather, the diffracted wave height becomes relatively large because the VLFS mooring position is not located right behind the breakwater line where the wave-reduction rate by wave diffraction due to the breakwater becomes the most effective. Therefore the vertical or composite breakwaters are generally suitable as a wave barrier in protecting the VLFS mooring area. In the following section, the acting wave force to vertical breakwater is explained.

Figure 6.3(4) shows the modified vertical breakwater developed with different purposes in mind. The modification has been done by improving the type and material of the original upright section in (4a). An upright wall with block masonry (4b) was popular in relatively shallow water areas. Cellular blocks (4c) were also employed to form the upright section. The invention caissons (4d) have been widely used because of reliability and low cost materials. The caisson breakwater with sloping top (4e) has been developed to reduce the horizontal component of wave force and to make more slender caissons. The perforated walls (4f) are employed mainly to reduce the size of reflected waves.

6.3   Wave-pressure formula

6.3.1   Goda formula and its evaluation

The evaluation of the wave pressure acting on a breakwater is important for determining the weight and width of the caisson in a composite breakwater. The Goda's formula (Goda 1974) is widely employed in the design of a breakwater. In 1973, Goda used his own theoretical and laboratory studies to establish a comprehensive formula for calculating the design wave force. After further modifications to account for the effect of oblique incident wave, the final formula was successfully applied to the design of vertical breakwaters that were built in Japan. The original Goda formula (Goda 1973) has many advantageous features, the main ones being:

1   It can be employed for all wave conditions including both standing and breaking waves.

2 The formula's design wave is the maximum wave height and can be evaluated by the given diagrams and/or equation.

3 It is based partially on nonlinear wave theory and can represent wave-pressure characteristics by considering two pressure components: the breaking and the slowly varying pressure components. Consequently, it is relatively easy to extend the Goda formula in order to apply it to other vertical-wall type structures.

4 The Goda formula clarifies the concept of uplift pressure on the caisson bottom, since the buoyancy of the upright section in still water and its uplift pressure due to wave action are defined separately. The distribution of uplift pressure has a triangular shape.

The Goda formula was subsequently extended to include the following parameters:

a the incident wave direction (Tanimoto et al. 1976);
b modification factors applicable to other types or vertical walls;
c the impulsive pressure coefficient (Takahashi et al. 1994b).

Recent collapses of breakwaters and seawalls are generated mainly in the presence of impulsive wave pressures (Hirayama et al. 2005). Therefore the consideration of impulsive wave pressure becomes very important and the design of breakwater should be modified when there is a large impulsive wave force.

In the extended Goda formula, the wave pressure acting along a vertical wall is assumed to have a trapezoidal distribution for both above and below the still-water level, while the upright section is assumed to have a triangular distribution as shown in Figure 6.4. The buoyancy is calculated using the displacement volume of the upright section in still water at the

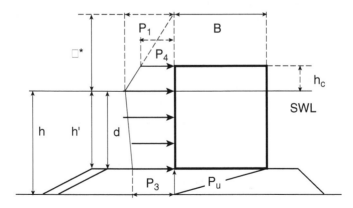

Figure 6.4 Goda pressure formula's parameter.

design-water level. As indicated, h denotes the water depth in front of the breakwater and d the depth above the armor layer of the rubble-mound foundation. Note that the rubble-mound foundation should be protected in armor-like unit-horizontal plate blocks so as to prevent displacement or scouring. The symbol h' indicates the distance from the design water level to the bottom of the upright section, and h_c the elevation of the breakwater above the design-water level. The elevation at which the wave pressure is extended, η^* and the representative wave pressure intensities p_1, p_3, p_4 and p_u can be written in a generalized form as

$$\eta^* = 0.75(1 + \cos\theta)\lambda_1 H_D$$
$$p_1 = 0.5(1 + \cos\theta)(\lambda_1\alpha_1 + \lambda_2\alpha^*\cos^2\theta)\rho g H_D$$
$$p_3 = \alpha_3 p_1 \tag{6.3}$$
$$p_4 = \alpha_4 p_1$$
$$p_u = 0.5(1 + \cos\theta)\lambda_3\alpha_1\alpha_3\rho g H_D$$

in which

$$\alpha_1 = 0.6 + 0.5\{(4\pi h/L_D)/\sinh(4\pi h/L_D)\}^2$$
$$\alpha^* = \max\{\alpha_2, \alpha_1\}$$
$$\alpha_2 = \min\{(1 - d/h_b)(H_D/d)^2/3, 2d/H_D\} \tag{6.4}$$
$$\alpha_3 = 1 - (h'/h)\{1 - 1/\cosh(2\pi h/L_D)\}$$
$$\alpha_4 = 1 - h_c^*/\eta^*$$
$$h_c^* = \min\{\eta^*, h_c\}$$

where θ is the angle between the direction of wave approach and a line normal to the breakwater; $\lambda_1, \lambda_2, \lambda_3$ the modification factors dependent on the structural type; H_D, L_D the wave height and wave length applied to calculate the design wave forces; α_1 the impulsive pressure coefficient; ρ the density of sea water; g the gravitational acceleration; h_b the offshore water depth at a distance five times the significant wave height $H_{1/3}$; and $\min\{a, b\}$ means the minimum of either a and b and $\max\{a, b\}$ the maximum of a and b.

6.3.2 Pressure component and pressure coefficients α_1, α_2 and α_I

Figure 6.5 shows the transition of wave pressure from non-breaking to impulsive pressure, where the pressure component is indicated by coefficients α_1, α_2 and α_I. The coefficient α_1 represents the slowly varying pressure component and α_2 the breaking pressure component, while α_I represents the impulsive pressure component, which includes the dynamic response effect on the caisson sliding. The coefficient α_I increases from 0 to 1.1 as the relative

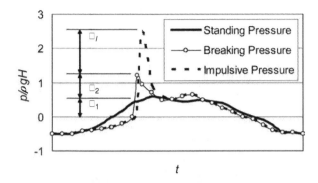

Figure 6.5 Image of generation of impulsive wave pressure.

depth decreases, and α_2 increases from 0 to 1.0, peaks, and then decreases as d/h_b decreases.

The coefficient α_I was obtained by reanalyzing the results of comprehensive sliding tests (Takahashi et al. 1994a). In the sliding test, a non-dimensional value representing the impulsive pressure component included the slowly varying component. The coefficient sloping α_I represents the effect of such slowly varying pressure components. The effect of the dynamic (impulsive) pressure by α_2 in Goda's formula does not under all conditions accurately estimate the effective pressure (equivalent static pressure) due to impulsive pressure. Therefore α_I was introduced.

Figure 6.6 shows a calculation diagram for α_I, in which it is expressed by the product of α_{I0} and α_{I1}, where α_{I0} represents the effect of wave height on the mound, that is,

$$\alpha_I = \alpha_{I0}\alpha_{I1}$$
$$\alpha_{I0} = H/d \quad \text{for } H \leq 2d \tag{6.5}$$
$$\alpha_{I0} = 2 \qquad \text{for } H > 2d$$

and α_{I1} represents the effect of the mound shape as shown by the contour lines. This can be evaluated using

$$
\begin{aligned}
\alpha_{I1} &= \cos \delta_2 / \cosh \delta_1 && \text{for } \delta_2 \leq 0 \\
\alpha_{I1} &= 1/\{\cosh \delta_1 (\cosh \delta_2)^{0.5}\} && \text{for } \delta_2 < 0 \\[4pt]
\delta_1 &= 20\delta_{11} && \text{for } \delta_{11} \leq 0 \\
\delta_1 &= 15\delta_{11} && \text{for } \delta_{11} > 0 \\[4pt]
\delta_2 &= 4.9\delta_{22} && \text{for } \delta_{22} \leq 0 \\
\delta_2 &= 3\delta_{22} && \text{for } \delta_{22} > 0
\end{aligned}
\tag{6.6}
$$

$$\delta_{11} 0.93(B_M/L - 0.12) + 0.36\{(h - d)/h - 0.6\}$$
$$\delta_{22} - 0.36(B_M/L - 0.12) + 0.93\{(h - d)/h - 0.6\}$$

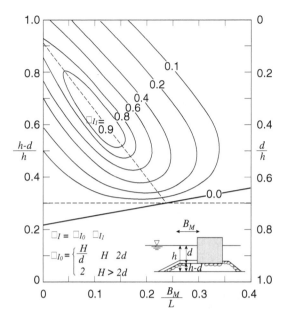

Figure 6.6 Calculation diagram of impulsive pressure coefficients.

Source: Takahashi et al. (1994a).

The value of α_I reaches a maximum of 2 at $B_M/L = 0.12, d/h = 0.4$ and $H/d > 2$. When $d/h > 0.7$, α_I is always close to zero and is less than α_2. It should be noted that the impulsive pressure significantly decreases when the angle of incidence θ is oblique.

For the ordinary vertical breakwater, $\lambda_1, \lambda_2, \lambda_3$ are taken as unity since the Goda formula was originally proposed for describing this type of breakwater. The modification factor λ_1 represents the reduction or increase of the wave's slowly varying pressure component, λ_2 represents the changes in the breaking-pressure component (dynamic-pressure component or impulsive-pressure component), and λ_3 represents the changes in the uplift pressure. These modification factors are employed for the other types of caisson breakwaters.

6.4 Wave-transformation

6.4.1 Wave-transmission coefficient

Incident waves are reflected at the breakwater and transmitted into the rear side of breakwater. The transmitted wave height and period become important when considering the stability of a VLFS moored in a water area sheltered by breakwaters. When the breakwater becomes relatively long as compared with the length of the VLFS, the influence of diffracted waves is negated and the wave transformation due to breakwater can be regarded as a two-dimensional (2D) phenomenon. Transmitted waves include waves that have overflowed as well as waves that have permeated through the rubble-mound breakwater or the foundation mound of composite breakwater. Recently, several breakwaters have been built with caisson (which were originally not permeable) having through-holes in order to enhance the exchange of the seawater within a harbor. In this case, it is necessary to examine the value of the wave-transmission coefficient, because the coefficient serves as an indicator of efficiency of the exchange of seawater.

Goda (1969) proposed a diagram to evaluate the transmitted wave height when they overflow a composite breakwater or permeate through a foundation mound. Figure 6.7 is used to calculate the transmission coefficient of composite breakwater. Even when the waves are irregular, the transmission coefficient agrees rather well with that shown in Figure 6.7. It has also been shown that Figure 6.7 is valid not only for the significant wave height, but also for the highest one-tenth wave height and the mean wave height.

For a porous, water-permeable structure such as a rubble-mound breakwater or a wave-dissipating concrete-block-type breakwater, Kondo's theoretical analysis (Kondo and Takeda 1983) may be employed. The following empirical equation may be used to obtain the transmission coefficient of a typical structure:

$$K_T = \frac{1}{1 + k_t\sqrt{H/L}} \quad \text{for stone breakwater} \tag{6.7}$$

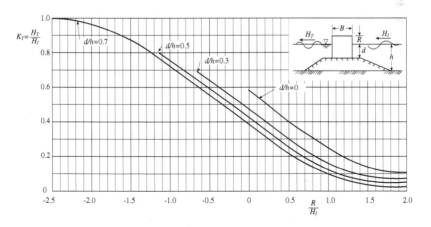

Figure 6.7 Graph for calculating the wave-height-transmission coefficient.
Source: Goda (1969).

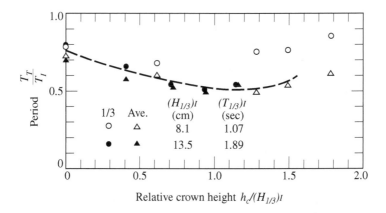

Figure 6.8 Variation of wave period in transmission at breakwater.
Source: Goda et al. (1974).

where $k_t = 1.26(B/d)^{0.67}$, B is the crown width of the structure; d is the depth from the water surface to the ground surface of the structure; H is the height of incident waves; and L is the wave length of transmitted waves. For a curtain-wall breakwater, the empirical solutions of Morihira et al. (1964) may be used.

The period of the transmitted waves drops to about 50–80% of the corresponding incident wave period. Figure 6.8 shows the variation of the transmitted wave period obtained in a 2D wave-flume test.

6.4.2   Wave-height distribution behind breakwater

Experimental set-up and calculations

In a harbor, the mooring area for a VLFS may have an influence on the diffracted waves as well as the transmitted waves. The represented wave condition is obtained by employing the wave-diffraction calculation and the estimation of transmitted waves. An experimental validation estimating the composed wave height is as follows: Figure 6.9 shows the plan view of the experimental basin for the distribution of wave heights in a harbor. The maximum water depth and width of the basin are 1 and 20 m, respectively. A directional snake-type wave generator with 30 segments, each 50 cm wide is installed in the basin. The hydraulic model for the harbor-wave distribution with scale of 1/100 was carried out in port K as shown in Figure 6.9. In the experiment, we employed a new type of breakwater with low crown-slope caissons that were placed in the location indicated by "slope-caisson breakwater." The wave-height distribution in

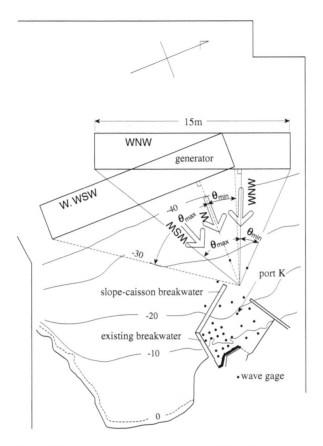

Figure 6.9 Experimental set-up to wave height inside harbor.

the port surrounded by the new and present breakwaters was measured at 29 locations with wave gauges. Detail configuration is indicated in Hiraishi (1994).

The principal direction of the generated waves varies in the three cases. Figure 6.10 shows the cross-section of a target breakwater with a low-crown slope caisson for the offshore breakwater. The slope-type breakwater has the ability to reduce horizontal wave forces acting on the caisson. Meanwhile the overtopping wave becomes significant for this type of breakwater. Therefore, the consideration of an overtopping wave is necessary in the analysis and design. The slope-type caisson is applicable to an isolated small harbor where a normal-type caisson is difficult to construct. The wave-transmission rate of the slope caisson becomes larger than that of the normal type.

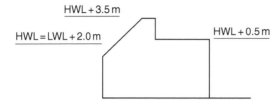

Figure 6.10 Example of slope caisson.

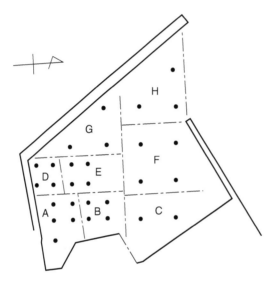

Figure 6.11 Zoning inside harbor for representative wave height.

Figure 6.11 shows the location of wave gauges in the target harbor. Black circles in Figure 6.11 represent the position of wave gages. Representative wave heights in each zone indicated in the figure were compared with the averaged wave height derived from calculations. For the calculations, the energy-balance equation method which has been improved to include the dissipation by wave breaking (Takayama et al. 1991) is employed in the computation of refraction and shoaling. The Takayama's diffraction model that assumes a constant water depth (Takayama 1985) is employed to evaluate the diffraction coefficient in a port. The wave height in the port can be calculated as the product of that computed in the energy-equivalent model and the diffraction coefficient. This method is one of the practical methods that are widely used for the design of harbors in Japan.

For the consideration of the overtopping wave, we employ the linear superposition of the direct-diffraction wave height and that of the overtopping waves. That is to say, the diffraction coefficient K_{D1} is computed with the assumption that there is no overtopping wave at the first stage. At the second stage, the diffraction coefficient K_{D2} is calculated with the temporary entrance at the slope-caisson breakwater with the assumption of no incident wave from the original harbor entrance. The period of the overtopping wave is assumed to be one-half of that of the incident wave (Hirakuchi et al. 1991). The height of the overtopping wave is assumed as the value of HK_T, where H and K_T indicate the incident wave height and the assumed transmission coefficient by the overtopping wave. The composed diffraction coefficient K_D can be computed by

$$K_D = \sqrt{K_{D1}{}^2 + K_{D2}{}^2} \tag{6.8}$$

Evaluation of computational method

For the validation of the numerical model, we usually carry out a hydraulic model test in a directional-wave basin. The directional-spreading function at the harbor entrance may differ from the target in the external effective generation area. For the calculation of harbor-wave heights, we also have to employ the distorted directional spreading when the harbor entrance is located slightly outside the effective test area. We propose the application of a simplified directional-spreading function for the input data to the diffraction test. The parameters θ_{min} and θ_{max}, as shown in Figure 6.9, express the minimum and maximum angular components for directional function, respectively.

The angular component with a propagating direction that may reach the target point (harbor entrance) can be limited because of the finite width of the generator. When θ_{min} and θ_{max} are employed as the minimum and maximum propagating angles, respectively, the distribution of directional waves with $S_{max} = 10$ should be modified. Figure 6.12 shows the modified directional-spreading function $G(f_p)$ (Mitsuyasu et al. 1975). The modified distribution indicated by the solid line in Figure 6.12 shows a higher and narrower distribution than the original direction-spreading function of $S_{max} = 10$ as indicated by the broken line.

Figure 6.13 shows the comparison of the measured and the estimated wave-height distributions in the target harbor. Figure 6.13(a) and (b) corresponds to the case of wave propagating direction WNW and W as shown in Figure 6.9, respectively. In the case of WNW direction, the overtopping was rather small in the experiment. The calculated wave height with different wave-transmission rate by overtopping K_T shows the same distribution. The estimated wave-height distribution agrees well with the measured

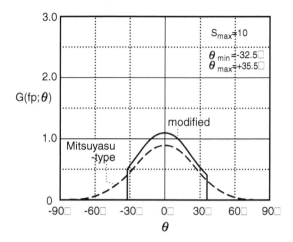

Figure 6.12 Modified-directional-spreading function.

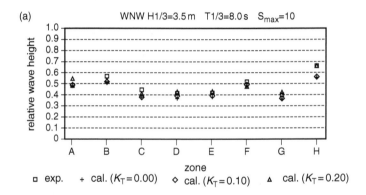

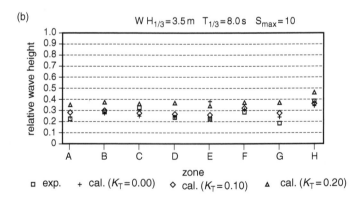

Figure 6.13 Comparison of measured and estimated wave height in harbor. (a) In case of
<WNW>. (b) In case of <W>.

result as indicated by the white markers. In the case of the W direction, the overtopping waves become larger so that the estimated wave heights with the various values of K_T express the different variations. By comparing the estimated wave-height distribution with the measured result, the computed results for $K_T = 0.1$ agree better with the results obtained experimentally than the other cases. Therefore, we can deduce that the overtopping transformation coefficient of the slope caisson is 0.1.

Based on the experiments, the following conclusions may be drawn:

1   The modified angular-spreading function with the angular limitation determined in the location and the width of the generated can be used for approximating the distorted angular-spreading function.
2   The harbor wave height is calculated as the linear superposition of diffracted and transmitted wave heights.

6.5   Example of experiment

In this section, we present the experimental investigation of the VLFS motion behind the breakwaters.

6.5.1   Experimental set-up

Figure 6.14 shows the arrangement of the dual-face directional-wave maker which was developed to expand the effective test area (Hiraishi et al. 1991) to a wider region in the basin and the VLFS model. One face of the wave maker is installed along the x-direction and another in the y-direction. The centerline of the VLFS is set to be parallel to the y-direction. The straight breakwater is installed along the y-direction to reduce wave heights acting on the floater. The dimensions are indicated on the model and the wave angle

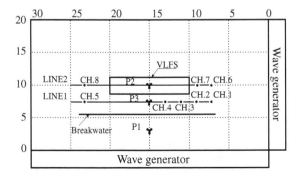

Figure 6.14 Experimental set-up for VLFS motion.

that is defined as the normal from y-direction is $0°$. The experimental wave and model conditions are as follows:

Water depth D: 50 cm (constant)
Significant wave height $H_{1/3}$: 2.0–10.0 cm
Significant wave period $T_{1/3}$: 1.05 s
Principal wave direction θ_p: $0°$ and $30°$
Angular spreading parameter S_{max}: 10, 50, 999 (unidirectional wave)
Spectrum type: JONSWAP
VLFS dimensions: length $L = 10$ m, width $B = 2$ m, height $D = 0.07$ m, draft $d = 0.02$ m
Stiffness of VLFS (EI/B): 765 Nm
Breakwater crown height h_c: 3.0 and 4.0 cm

Figure 6.15 shows the position of the mooring system for the VLFS model. Four mooring joints with universal joints are employed. In the experiment, three components (surge in the x-direction, sway in the y-direction, yaw in the horizontal face) are measured and analyzed. The detail of this experiment is indicated in Hiraishi et al. (2000).

6.5.2 Wave characteristic around breakwater

Figure 6.16 shows the wave-directional spectrum measured in a wave gauge array located at P.2. The relative crown height h_c/H_i is 0.43 and the directional-spreading parameter $S_{max} = 10$. Figure 6.16(a) and (b) show the 2D directional function which was calculated by integrating the directional-wave spectrum in the wave-frequency domain for the cases of $\theta_p = 0°$ and $30°$, respectively. In the figures, the directional function measured in the center of VLFS's location is indicated for cases with and without a breakwater. The wave energy behind the breakwater becomes very small when the principal wave direction θ_p becomes 0 as shown in Figure 6.16(a). However

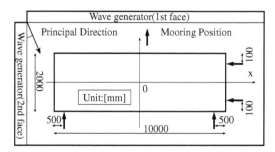

Figure 6.15 Mooring system for the VLFS.

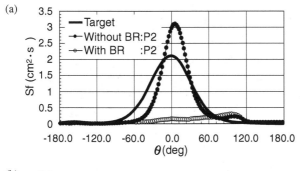

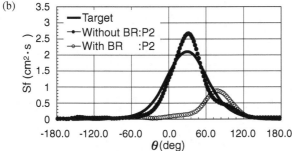

Figure 6.16 Variations of wave-directional spectrum by breakwater. (a) $\theta_p = 0°$, (b) $\theta_p = 30°$.

a small peak is generated at $\theta = 90°$ which may be caused by overtopping waves. In the oblique-wave case in Figure 6.16(b), the wave energy behind the breakwater has a strong peak at $\theta = 80°$ which is caused from the obliquely proceeding wave components. In the oblique-wave cases, the influence of overtopping wave becomes relatively small when compared with the normal incident wave because the overtopping wave rate decreases as the wave angle increases.

The wave-transmission coefficient was calculated by using the composed wave height that was experimentally obtained. Equation (6.9) is used for estimating the wave-transmission coefficient at the breakwater.

$$K_t^* = \sqrt{K_t^2 - K_{t(2\ \text{cm})}^2}$$
$$K_t = H_{\text{exp}}/H_{\text{in}}$$
$$K_{t(2\ \text{cm})} = H_{\text{exp}(2\ \text{cm})}/H_{\text{in}(2\ \text{cm})}$$

(6.9)

where H_{exp} is the measured wave height; H_{in} the incident wave height; K_t^* the wave transmission coefficient by overtopping wave; and $K_t = 2$ cm is the

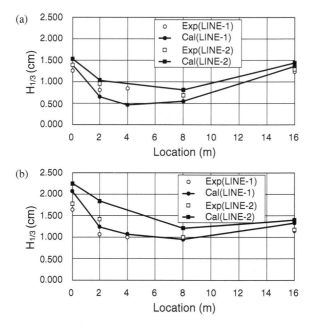

Figure 6.17 Comparison of measured and calculated diffracted wave heights behind breakwater. (a) $\theta_p = 0°$, (b) $\theta_p = 30°$.

measured wave height for the case of $H_{in} = 2$ cm. We confirmed that no wave overtopping occurs at this level.

Figure 6.17 shows the comparison of measured wave heights and the calculated wave heights according to the Takayama wave-diffraction model for the case of $H_{in} = 2$ cm. The measured wave-height distributions are in good agreement with the calculated results. Therefore the measured wave height is composed of only the diffracted waves including the wave energy directly influenced as oblique waves.

Figure 6.18 shows the variation of wave-transmission coefficient for the relative breakwater crown height. The incident wave is the directional waves with energy spreading $S_{max} = 10$ for the normal incident ($\theta = 0°$) and oblique incident ($\theta = 30°$). The rigid line shows the 2D experimental result obtained by Kondo and Sato (1963). The wave-transmission coefficient for the directional waves becomes relatively small when compared with the unidirectional wave condition as shown in Figure 6.18. The wave-transmission coefficient becomes almost 0 when the relative crown height becomes larger than 0.7 times the incident wave height for both cases of the normal and oblique incidents. The diffracted wave height behind the breakwater becomes slightly larger in the oblique incidents than the normal incidents.

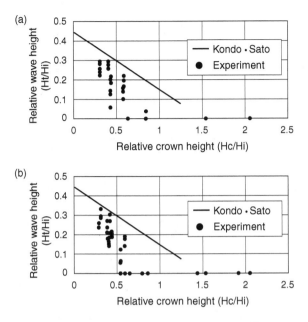

Figure 6.18 Variations of wave-transmission coefficient with respect to crown height, (a) $\theta_p = 0°$, (b) $\theta_p = 30°$.

The diffracted wave heights increases as the wave angle increases while the transmission wave height decreases as the wave angle increases. Therefore, the composed wave height may not be remarkably influenced even if the wave direction changes.

6.5.3 Results of measured VLFS motion

In this section, the motion of the VLFS is demonstrated for cases with breakwater and without breakwater.

VLFS motion without breakwater

Figures 6.19 and 6.20 show the variations of surge and yaw-motion amplitudes when the wave direction varies from 0° to 90° and the directional parameter S_{max} changes from 10 to 999 for the same wave height ($H_{1/3} = 5$ cm) and period ($T_{1/3} = 1.05$ s). The symbol S_f represents the incident wave spectrum. In Figure 6.19, the surge motion spectral density increases as the θ_p increases from 0° to 90°. In the case of the constant wave direction ($\theta_p = 90°$), the surge amplitude becomes larger in the case of unidirectional waves ($S_{max} = 999$) than in case of directional waves

(a)
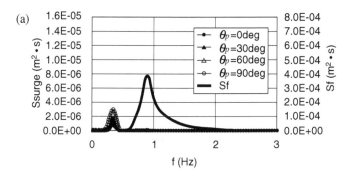

(b)
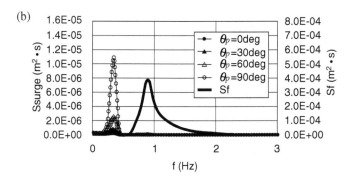

(c)
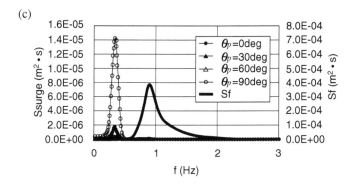

Figure 6.19 Variation of surge amplitude. (a) $S_{max} = 10$, (b) $S_{max} = 50$, (c) $S_{max} = 999$ (uni-directional).

$(S_{max} = 10, 50)$. In the cases of $\theta_p = 0°$ to $60°$, the surge amplitude becomes larger in the directional waves than in the uni-directional waves.

In Figure 6.20, the yaw amplitude becomes large in the cases of $\theta_p = 0°$ and $30°$. For different S_{max} cases, the yaw amplitude becomes larger for the

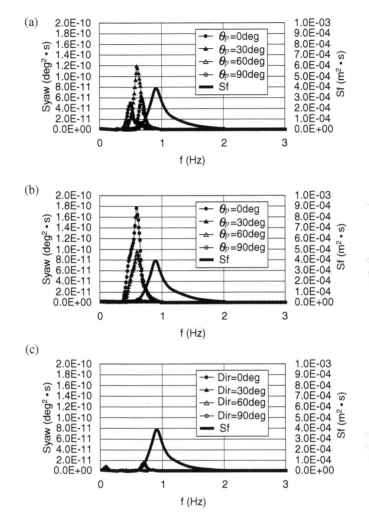

Figure 6.20 Variation of yaw amplitude. (a) $S_{max} = 10$, (b) $S_{max} = 50$, (c) $S_{max} = 999$ (uni-directional).

directional waves than for the unidirectional waves. Meanwhile the peak period of surge and sway motion differ from the peak period of incident waves in all cases. The natural oscillation frequencies of target VLFS derived from the free-oscillation test are given in Table 6.1. The peak frequency of surge and sway observed in Figures 6.19 and 6.20 are almost the same as the natural frequency shown in Table 6.1 but it is shorter than the incident wave's one. Therefore the amplitude of surge and sway of a VLFS moored offshore becomes larger in the directional waves than in the uni-directional waves at specified wave angles. Meanwhile the amplitude of the yaw motion becomes larger in directional waves for all wave directions. The directional characteristics in target area are important when determining the VLFS motion.

VLFS motion with breakwater

Figure 6.21 shows the spectrum of yaw motion in the case for breakwater of $h_c = 4.0$ cm. The incident wave condition is $\theta_p = 30°$, $S_{max} = 10$ and $T_{1/3} = 1.05$ s, $H_{1/3} = 2.0$–10.0 cm. Thus the relative crown height h_c/H_{in} ranges from 0.4 to 2.0. For the case of $H_{1/3} = 10.0$ cm (i.e., $h_c/H_{in} = 0.4$), the yaw amplitude becomes large because of the influence of overtopping wave is significant. Figures 6.22 and 6.23 show the spectral motion amplitudes of VLFS moored behind the breakwater. The wave condition is defined by

Significant wave height $H_{1/3} = 5.0$ cm ($h_c/H_{in} = 0.8$, no wave overtopping)
Significant wave period $T_{1/3} = 1.05$ s
Directional spreading $S_{max} = 10$
Wave directions $\theta_p = 0°$ (see Figure 6.22) and $\theta_p = 30°$ (see Figure 6.23)

Table 6.1 Natural oscillation frequencies of VLFS

Motion component	Natural oscillation frequency (f)
Surge	0.35
Sway	0.34
Yaw	0.52

Source: Hiraishi et al. (2000).

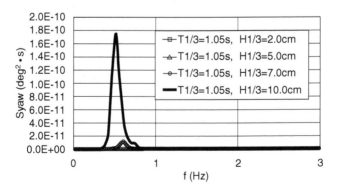

Figure 6.21 Relation of yaw amplitude and h_c/H_i.

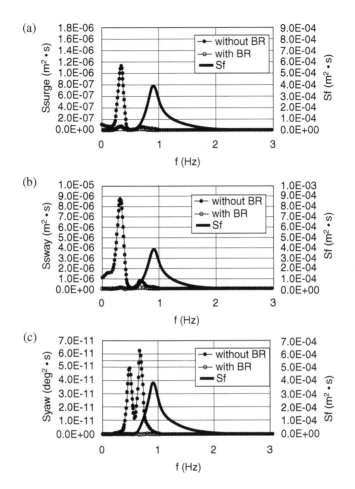

Figure 6.22 Reduction of VLFS motion due to breakwater ($\theta_p = 0°$). (a) Surge, (b) Sway, (c) Yaw.

In Figure 6.22, the motion amplitude of each motion components becomes significantly small after the construction of breakwater. In case of $\theta_p = 30°$ (Figure 6.23), the reduction ratio of the motion amplitude becomes smaller than in the case of $\theta_p = 0°$. The reason for the smaller reduction of the motion amplitude may be due to the influences of the diffracted oblique waves from the breakwater edge. In the case of no wave overtopping, the reduction rate of spectral-peak density become less than 0.1 in all cases.

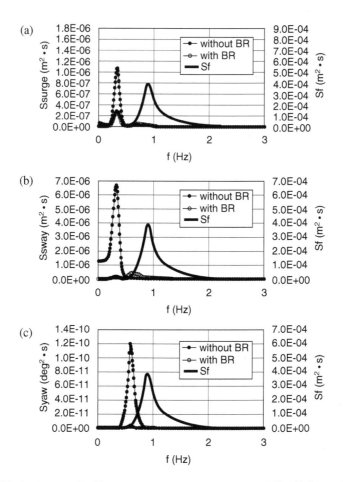

Figure 6.23 Reduction of VLFS motion due to breakwater ($\theta_p = 30°$). (a) Surge, (b) Sway, (c) Yaw.

Based on the VLFS experiment, the following conclusions may be drawn:

1   The wave-transmission coefficient due to overtopping wave becomes smaller in the directional waves when compared with their uni-directional counterparts. The transmission coefficient is almost 0 when the crown height is higher than 0.7 times the incident wave height.

2   The VLFS motion amplitude due to the directional waves did not show any significant changes even when the wave direction was varied.

3   The motion of a VLFS moored behind breakwater becomes very small for the case where the crown height is higher than 0.6 H_{in}.

References

Goda, Y. (1969) Re-analysis of laboratory data on wave transmission over breakwaters, *Report of Port and Harbour Research Institute*, 8(3), pp. 3–18.

Goda, Y. (1973) A new method of wave pressure calculation for the design of composite breakwater, *Report of Port and Harbour Research Institute*, 12(3), pp. 3–29.

Goda, Y. (1974) New wave pressure formulae for composite breakwaters, *Proceedings of 14th International Conference on Coastal Engineering*, pp. 1702–1720.

Goda, Y., Suzuki, Y., and Kishira, Y. (1974) Some experiences in laboratory experiments with irregular waves, *Proceedings of 21st Japanese Conference on Coastal Engineering*, pp. 237–242.

Hiraishi, T. (1994) Laboratory measurement of directional random wave heights in a harbor, *Technical Note of the Port and Harbour Research Institute*, No. 788, p. 32.

Hiraishi, T., Mansard, E.P.D., Miles, M.D., Funke, E.R., and Issacsons, M. (1991) Numerical and experimental validation for a diffraction model of directional wave generation, *Technical Report*, TR-HY-037, NRC, p. 133.

Hiraishi, T., Hirayama, K., Maruyama, H., Kato, S., Saito, Y., and Namba, Y. (2000) Motion of elastic large floater behind offshore breakwater, *Proceedings of 47th Japanese Conference of Coastal Engineering*, pp. 831–835.

Hirakuchi, T., Ikeno, M., Ohashi, H., Kashiwagi, H., and Higuchi, T.M. (1991) Characteristics of wave height, period and spectrum of overtopping wave at breakwater, *Proceedings of 38th Japanese Conference on Coastal Engineering*, pp. 506–510.

Hirayama, K., Hiraishi, T., Minami, Y., Okuno, M., and Minemura, K. (2005) Disaster pattern induced by storm waves in 2004, *Annual Journal of Coastal Engineering*, JACE, pp. 1316–1320.

Kondo, T. and Sato, I. (1963) Study on crown height of breakwaters, *Monthly Journal of Civil Engineering Research*, Institute for cold region, No. 117, pp. 1–15.

Kondo, T. and Takeda, H. (1983) Wave absorbing structures, ISBN 4-627-49060-7, Morikita Ltd., pp. 70–129.

Mitsuyasu, H., Tasai, T., Mizuno, S., Ohkusu, M., Honda, T., and Rikiishi, K. (1975) Observation of the directional spectrum of ocean wave using a cloverleaf buoy, *J. Physical Oceanography*, 5(4), pp. 750–760.

Morihira, M., Kakizaki, S., and Goda, Y. (1964) Experimental investigation of a curtain-wall breakwater, *Report of Port and Harbour Research Institute*, 3(1), p. 27.

Nagai, T. and Nukata, K. (2004) Recent improvement of real-time coastal wave-tide information system in Japan, *Proceedings of ACECC-TCI Workshop (Wave, Tide Observations and Modeling in the Asia-Pacific Region)*, pp. 1–12.

Takada, E., Morohoshi, K., Hiraishi, T., Nagai, T., and Takemura, S. (2002) Distribution of the wave, storm surge and tsunami design conditions on Japanese nationwide coastal structures, *Technical Note of National Institute for Land and Infrastructure Management*, No. 88, p. 132.

Takahashi, S. (1996) Design of vertical breakwaters, Reference Document No. 34, *Port and Harbour Research Institute*, Ministry of Transport, p. 85.

Takahashi, S., Tanimoto, K., and Shimosako, K. (1994a) Experimental study on impulsive pressures on composite breakwaters, *Report of Port and Harbour Research Institute*, 31(5), pp. 33–72.

Takahashi, S., Tanimoto, K., and Shimosako, K. (1994b) A proposal of impulsive pressure coefficient for design of composite breakwaters, *Proceedings of International Conference on Hydro-technical Engineering for Port and Harbour Construction*, pp. 20–30.

Takayama, T. (1985) Computation of wave height distribution inside a harbor, *Proc. Conf. Numerical and Hydraulic Modeling of Ports and Habours*, pp. 295–302.

Takayama, T., Ikeda, N., and Hiraishi, T. (1991) Practical computation method of directional random wave transformation, *Report of Port and Harbour Research Institute*, 30(1), pp. 21–67.

Tanimoto, K., Moto, K., Ishizuka, S., and Goda, Y. (1976) An investigation on design wave force formulae of composite-type breakwaters, *Proceedings of 23rd Japanese Conference on Coastal Engineering*, pp. 11–16.

The WAMDI Group (1988) The WAM model—A third generation ocean wave prediction model, *J. Phys. Oceanography*, 18, pp. 1775–1810.

Chapter 7

Model experiments for VLFS

Shigeo Ohmatsu

7.1 Introduction

It is common practice to conduct a model experiment in a water basin in order to investigate the behavior of a vessel or a marine structure in waves. Although the accuracy of estimating the dynamic characteristics of a vessel or a marine structure by the theoretical analysis and numerical simulation has been improved based on the recent development of electronic computers, the model experiments of vessels and marine structures have been conducted over 100 years and new phenomena are often discovered by such model experiments even in these modern times. In particular, for a very large floating structure (VLFS) that has never been built, even if the reliability of the theoretical analysis is very high, it is crucial to validate the theoretical analysis or investigate the behavior by a model experiment of VLFS (see Figure 7.1).

Basically, there are three types of objectives in conducting model experiments of floating structures (SNAJ 1997). First to validate the result of the theoretical analysis; that is to assess how correct the analysis is. The oscillation-response experiment of VLFS in regular waves of small wave height falls in this objective category. Second, to investigate matters that

Figure 7.1 Model experiments in wave tank of conventional vessel (left) and VLFS (right).

are difficult to estimate theoretically or analyze. The investigation of fluid forces due to the fluid viscosity or the study of the nonlinear effects of wave heights to the behavior of a VLFS in large wave heights is in this category. Third to investigate the overall performance characteristics of a VLFS in a realistic situation as far as possible. By assuming that VLFS is moored to several mooring devices, the experiment to show how the VLFS will behave in multidirectional irregular waves falls into this category. The construction of a VLFS requires that the floating unit modules be joined together on the ocean. A model experiment of a VLFS on the joining operation of multifloating modules falls into this category.

Though model experiments of a VLFS are very important, there are peculiar problems in these model experiments. The first problem is the fact that an actual VLFS is, by definition, "very large." In order to conduct a model experiment of a large floating structure in an available water basin, it is necessary to reduce the scale of the geometrically similar model of a large floating structure to an extraordinarily small scale. This may cause a problem in the generation capability of a water basin (i.e., how to accurately generate short waves) and the accuracy of measuring the behavior of the model.

The second problem is that the bending rigidity in the vertical direction is relatively small and that the response and deformation of a VLFS as an elastic body are quite large in the vertical direction. This is because the dimensions in the vertical direction of VLFS are extremely small compared to the ones in the horizontal directions as mentioned before. For an ordinary vessel, its deformation as an elastic body does not pose many problems. However, for VLFS, this "hydroelastic problem" is quite important and it should be taken into consideration for the estimation of a VLFS's behavior. In order to examine these phenomena in the model experiment, it is necessary to make the rigidity of the model similar to the rigidity of the actual VLFS. However, it is very difficult to make the rigidity of a small-scale model similar in order to meet the law of similarity.

This chapter describes how to conduct an accurate model experiment by overcoming these problems and then introduce some examples of model experiments of VLFS.

7.2   Model

There are some problems that one must be aware of when fabricating VLFS models for experiments. These problems include how to decide the scale and how to build the model. It is clear that the exterior dimensions of the actual VLFS must be reduced in accordance to the selected scale adopted for the model. In addition, it is also necessary to reduce the overall weight, weight distribution and rigidity distribution of the actual VLFS when conducting the model experiments in order to study its oscillation characteristics.

7.2.1 Law of similarity

First, the law of similarity in the model experiments for the hydroelasticity problems of VLFS is discussed. The nondimensional expressions of the following items under the similarity condition related to the elastic response of VLFS in waves (Endo 1991) are

- Geometrical parameters $(\lambda/L, \xi/L, \text{etc.})$
- Mass parameter $(M/\rho L^3)$
- Froude number $(V/\sqrt{gL})$
- Vertical bending rigidity parameter $(EI/\rho gL)$
- Shearing rigidity parameter $(GA/\rho gL^3)$
- Inertia of rotation of the cross-section parameter $(\psi/\rho L^4)$
- Structural damping coefficient (c)

where ρ is the water density; g the gravitational acceleration; λ the wave length; ξ the wave height; L the length of structure; V the velocity of fluid; EI the vertical bending rigidity; GA the shearing rigidity; and ψ the inertia of rotation of the cross-section. Items 4 through 7 are related to elasticity of the structure. Among them, 4 is a particularly important item and various technologies have been developed to make the bending rigidity of a model similar to that of a real VLFS. However, items 5 through 7 are often neglected by assuming that their effects are small. Thus, the law of the similarity for making a model VLFS becomes as listed in Table 7.1.

The appropriateness of the "law of similarity" for bending rigidity can be confirmed as follows. First, consider a static problem for obtaining the deformation of an elastic beam whose ends are supported and a uniform load w (such as self-weight) is applied on the beam as shown in Figure 7.2. Then the vertical displacement y of the beam can be expressed by

$$y = \frac{wl^3x}{24EI}\left(1 - \frac{2x^2}{l^2} + \frac{x^3}{l^3}\right) \tag{7.1}$$

Table 7.1 Law of similarity

Geometrical condition	$L_m = \alpha L_f$
Mass	$M_m = \alpha^3 M_f$
Time condition	$T_m = \sqrt{\alpha} T_f$
Rigidity	$EI_m = \alpha^5 EI_f$

Notes
α: scale ratio; m: model; f: full scale.

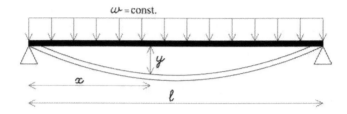

Figure 7.2 Deflection of beam under uniform load.

Figure 7.3 Elastic wave of VLFS due to regular waves.

Thus, if the scale of the displacement y is reduced by the scale ratio α, then the scale of EI should be the fifth power of α. This can also be verified by the dynamic elastic response of the VLFS in waves. The deflection of the VLFS in regular waves can be regarded approximately as wave phenomena similar to the progressive waves on the water surface (see Figure 7.3).

The wave number of this elastic wave is given by Ohmatsu (2005).

$$\frac{\omega^2}{g} = \left[1 - \left(\frac{\omega}{\omega_0}\right)^2 + \left(\frac{k}{k_p}\right)^4\right] k \tanh kh = k_w \tanh k_w h \qquad (7.2)$$

where

$$\omega_0 = \sqrt{\frac{\rho g}{m}} \qquad (7.3)$$

$$k_p = [\rho g/(EI/B)]^{1/4} \qquad (7.4)$$

h is the water depth; ω the wave frequency; k_w the wave number of water wave; B and m are the width and mass per unit area of the VLFS, respectively. If the wave number k of the elastic wave is to be made similar to that of the actual VLFS, it can be seen from Eq. (7.4) that EI of the model should be the fifth power of α of the actual flexural rigidity of VLFS.

7.2.2 Determination of scale ratio

Take for an example the model experiments for a floating airport. Careful attention should be paid to maintaining the measurement accuracy of the experiments. It would be necessary to use a far smaller scale model than the actual model because of the very large dimensions of the airport in the horizontal direction. When conducting a model experiment of an ordinary vessel or a floating structure, the scale of the model must be maintained to be about one hundredth (1/100), even for a small-scale model. Therefore, if the model experiments of a floating airport having the length of 2,250 m are to be conducted, the dimensions and the weight of the model should be set according to Table 7.2.

Since a scale ratio of 1/100 gives the total length of the model as 22.5 m, it is very difficult to conduct the experiments. A scale of 1/150 with a total length of 15 m is possibly the upper limit of a model experiment. In practice, the actual model experiments of VLFS are to be conducted using a much smaller scale. When a model of 1/500 or smaller scale is used, it becomes rather difficult to build a model to study the dynamic characteristics of the actual VLFS.

Further, the model limitation arises from the wave-making ability of the water basin, that is, how accurately the water basin can make short waves. The wave period of the significant waves in an actual sea lies in the range of 4–15 s. It is difficult to make waves in an ordinary water basin having the time scale of less than 1/20 in accordance with Froude's law of similarity.

In addition, the scale of a model should be considered from the viewpoint of model making, model transportation and model setting. For instance, when setting a model in a water basin, it would be necessary to investigate (a) whether to fabricate the model on land and move it into the water basin by transporting it on a rigid frame; (b) whether to fabricate the model in the water in the same way as an actual VLFS due to be fabricated under floating conditions; or (c) whether to fabricate the model in a dry water basin and then fill the basin with water after completing the model.

Table 7.2 Principal particulars of various scale models

	Prototype	1/150 model	1/300 model	1/500 model
Length scale α	1	1/150	1/300	1/500
Time scale $\sqrt{\alpha}$	1	1/12.25	1/17.32	1/22.36
Length	2,250 m	15 m	7.5 m	4.5 m
Breadth	450 m	3 m	1.5 m	0.9 m
Height	4.5 m	30 mm	15 mm	9 mm
Draft	1.0 m	6.7 mm	3.3 mm	2 mm
Displacement	1,038 t	300 kg	37.5 kg	8.1 kg
EI (Longitudinal)	2.25×10^{13} Nm^2	2.96×10^8 Nmm^2	9.26×10^6 Nmm^2	7.20×10^5 Nmm^2

7.2.3   Material and fabrication method

Once the scale ratio of a model is decided upon, the model should be fabricated into the scale that satisfies "the law of similarity." It is necessary to think of how to simultaneously satisfy the condition of weight and bending rigidity. The model is to be fabricated using various materials. Physical characteristics of typical materials are listed in Table 7.3.

Consider the design example of a 1/150-scale floating airport model using parameters in Table 7.2. The target weight of the model is 300 kg and the target bending rigidity EI is 2.96×10^8 Nmm².

Single plate

First, consider the fabrication of a uniform plate model as shown in Figure 7.4. The main material of the model should be aluminum with a sufficient weight and rigidity. A polyethylene foam is used as buoyancy material which has negligible weight and rigidity.

When using a 2.5-mm thick aluminum plate, then $I = \frac{t^3}{12}b = \frac{2.5^3}{12}3000$ mm⁴ and $E = 0.703 \times 10^5$ N/mm². Therefore, the flexural rigidity $EI = 2.75 \times 10^8$ Nmm². Its weight $W = L_{(m)} \times B_{(m)} \times t_{(m)} \times \gamma_{(t/m^3)} = 15 \times 3 \times 0.0025 \times 2.7 = 0.304$ ton. The rigidity and the weight almost satisfy the targeted figures.

Sandwich type

Next, consider making a sandwich-type structural model using material having a small Young's modulus as shown in Figure 7.5. The required rigidity is achieved by adjusting the height of the sandwich. An acrylic plate is to be used for the model. The shell of the model comprises 1-mm thick acrylic plates and its core is a 7-mm thick material whose rigidity is negligible.

Table 7.3 Relative density (γ) and Young's modulus (E) of various materials

	Steel	Aluminum	Acryl	FRP	Polyvinyl
γ	7.7	2.7	1.2	1.6	1.4
$E \times 10^{-5}$ (N/mm²)	2.06	0.703	0.03	0.12	0.025

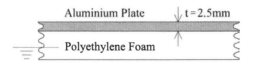

Figure 7.4 Cross-section of VLFS model (single plate).

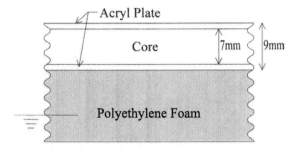

Figure 7.5 Cross section of VLFS model (sandwich type).

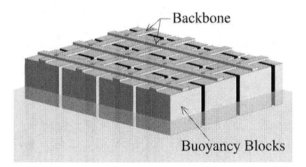

Figure 7.6 Backbone type model VLFS.

Based on the dimensions, $I = \frac{9^3 - 7^3}{12} 3000\,\text{mm}^4$ and the Young's modulus $E = 0.03 \times 10^5\,\text{N/mm}^2$, the flexural rigidity is $EI = 2.90 \times 10^8\,\text{Nmm}^2$. Its weight is $W = 15 \times 3 \times (0.002 \times 1.2 + 0.007 \times 0.6) = 0.297\,\text{ton}$. Thus, both targeted weight and rigidity are satisfied. In this case, the rigidity can be adjusted by changing the height of the sandwich without changing the weight.

Backbone

There is another method for fabricating a model that is by having many floating blocks and tying them up with the backbone as shown in Figure 7.6 (other than fabricating a uniform plate or a sandwich-type model as described above). By keeping clearances between blocks, a required rigidity is given only by the backbone.

For example, consider a model comprising 10×50 blocks. If a 4-mm thick and 80-mm wide aluminum bar is used for 10 backbones, then $I = \frac{4^3}{12} 80 \times 10\,\text{mm}^4$ and $E = 0.703 \times 10^5\,\text{N/mm}^2$. Therefore the flexural rigidity

Figure 7.7 Photo of backbone type model.

is $EI = 3.00 \times 10^8 \, \text{Nmm}^2$. Its weight is $W = 0.004 \times 0.08 \times (15 \times 10 + 3 \times 50) \times 2.7 = 0.259$ ton. Assuming that the remaining 41 kg represents the weight of 500 buoyancy blocks, the rigidity and weight of the model are both satisfied. Figure 7.7 shows an example of the fabricated backbone model. This method can be used to adjust the bending rigidity of a model by changing the backbones' longitudinal and transverse dimensions and the number of backbones, even though the longitudinal and transverse direction bending rigidities of the actual VLFS are different. However, it is necessary to prepare the model in many small blocks in order to obtain the smooth elastic response of the model but it is troublesome to make these blocks.

7.2.4 Mooring device

The mooring device of the VLFS model may be represented by a simple spring that prevents the drifting of the model when only the elastic response of the VLFS in waves is being considered. However, when examining the behavior of a moored VLFS in a horizontal plane it is necessary to use a mooring-device model that simulates the reaction characteristics of the actual mooring device. The spring constant of the model should be α^2 times of the actual VLFS's spring constant.

The dolphin-fender system is the typical mooring facility of a VLFS. The reaction forces of the fender exhibit nonlinear characteristics. Thus, a model of the mooring facility should be fabricated by a combination of multiple linear spring units in order to simulate accurately the nonlinear characteristics. An example of the reaction force of a mooring device is shown in Figure 7.8.

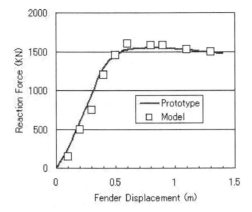

Figure 7.8 Reaction characteristics of model mooring device (in full scale).

When the water depth at the installation site is deep, it is common to use the catenary mooring method with many chain lines. In this case, the weight of the chain in the water should be adjusted so that it is similar in weight to the actual VLFS's chain lines.

7.3 Experimental basin

Vessels and marine structures have been commonly tested in water basins for investigating their seaworthiness in waves. Usually these water basins are equipped with wave-making devices. Model experiments of VLFSs are conducted in these water basins as well. Requirements for such water basins and wave-making devices peculiar for the model experiments of VLFSs are described below:

7.3.1 Water basin

During model experiments in a wave basin, it is well known that scattered waves of the model reflect at the basin walls and hit the model again. As a result, the complicated effects of the walls cause problems to the experiment. These effects of the walls are more significant in the experiment of a marine structure that does not have a forward speed than the experiment of an ordinary moving vessel. In order to avoid these effects, it is desirable to use a wide rectangular-shaped water basin. As a VLFS has large horizontal dimensions, it is desirable to use a water basin having a water surface area that is as large as possible.

On the other hand, it is desirable to have a shallow water depth. As a VLFS is usually located in shallow water, a water basin used in conducting a model experiment of a vessel is too deep for the VLFS model experiment. If the water

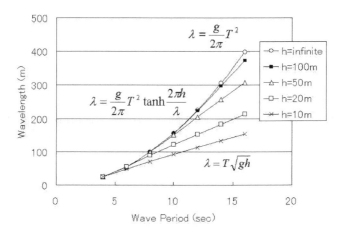

Figure 7.9 Relation of wave period and wave length for each water depth.

depth is set to be between 15 and 60 m, then the corresponding depth of the water basin for the 1/150 scale model experiment should be between 10 and 40 cm. As shown in Figure 7.9, the relationship between a wave period and wave length depends greatly on the water depth. Thus, it is desirable to conduct the model experiment in a water depth that is similar to the actual water depth where the VLFS is to be installed. If the depth of a water basin cannot be adjusted, a sufficiently wide temporary floor should be installed in the basin in order to make the water depth similar to the actual water depth.

As VLFS has no forward speed, it is not necessary to tow the model of the VLFS in the experimental basin. However, if a towing carriage is available in the basin, it would be convenient to use it for arranging measuring instrument units. Also it may be used to measure the current forces generated from towing the VLFS model.

7.3.2 Wave-making device

In general, square basins are equipped with wave-making devices on one or two sides. Regular waves or irregular waves having any frequency spectrum may be generated. Most of the wave-making devices are those that can make two-dimensional (2D) long-crested irregular waves. In order to simulate a condition that is similar to an actual sea surface, recent wave-making basins are equipped with devices that can simulate the directional spectrums by the multi-segmented wave makers as shown in Figure 7.10.

For the model experiments of a VLFS, the wave-making device should be able to generate accurately the waves of small lengths and small heights because the scale of the model is very small.

Figure 7.10 Photo of multi-segmented wave maker.

The significant wave periods in actual sea conditions are in the order of 3–15 s. When using a model of 1/150 scale, its time scale becomes approximately 1/12 and the wave period for the experiments is in the range of 0.24–1.22 s. In general, it is difficult to make waves having periods of less than 0.3 s. Moreover, when using a model of scale of 1/150, the equivalent a 1-m high real wave should be 6.7 mm high waves. It is clear that accurate generation of very small wave height is necessary.

As for the experiment investigating the horizontal motion of a moored VLFS, the slowly varying drifting forces caused by irregular waves are important. Thus, for a row of irregular waves, it is necessary to control the probability of a group of continuous high waves and the long period water surface elevation fluctuations. In this case, it is necessary to make waves by the wave-making signals that control the amplitude and phase difference of the component waves.

7.4   Measuring instrument

Quantities to be measured in the model experiments of a VLFS include the vertical displacements and the strains of the floating structure and the reaction force of the mooring devices.

In order to investigate the elastic response of the floating structure, it is necessary to measure accurately the vertical displacements and strains

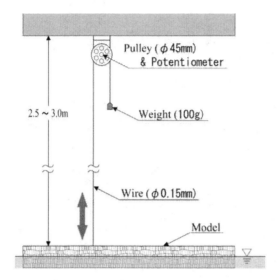

Figure 7.11 Potentiometer to detect the vertical displacement of VLFS.

at various points. However, as mentioned above, the displacement of the model is extremely small when compared to the horizontal dimensions and, moreover, the model is an easily deformable structure. Thus, a non-contact-type measuring instrument is desirable so that the instruments have no influence on the response of the model. Presently, laser displacement meters are put into practical use. However, it is very costly when used at many measurement points. Thus, the most practical method at present is to use a potentiometer as shown in Figure 7.11 (Yago 1996).

The potentiometer is a low-cost unit. It is accurate and does not disturb the behavior of the floating structure even if the instrument is used at many measurement points on the floating structure. It is also possible to obtain the elastic response from the measured result of the vertical acceleration that is set at many points.

The strains can be measured by strain gauges that are attached to the floating structure. The bending moments can be converted from the measured strains using the following equation

$$M' = \frac{\varepsilon \cdot EI'}{z/2} \tag{7.5}$$

where M' is the bending moment; EI' the bending rigidity in unit width and z the thickness of the strength member of the model.

The reaction forces acting on the mooring devices can be converted from the measured horizontal displacements of the mooring-device models.

7.5   Example of model experiments of VLFS

7.5.1   Elastic-response experiments in waves

As an example of the elastic-response-model experiments of a pontoon-type VLFS, the experiment that was jointly conducted by the Technological Research Association of Mega-Float and the National Maritime Research Institute is introduced (Ohkawa 2000).

Purpose of the experiment

The model experimental tests were conducted to validate the theoretical elastic-response analyses of a VLFS under the following conditions: mutual effect of a VLFS and a breakwater; rectangular VLFS of nonuniform rigidity; and nonrectangular VLFS of uniform rigidity.

Model

The length of the actual VLFS to be built was assumed to be 1,000 m. By adopting the scale of the models of 1/100, the length of the models was 10 m. Table 7.4 lists the dimensions of the models.

The three types of model were fabricated carefully in order to satisfy the experiment objectives (see Figure 7.12). M1 was a basic model having a structure of a 5-mm thick aluminum plate with a 65-mm thick polyethylene foam for buoyancy. The rigidity and the weight were adjusted based on the law of similarity. M2 was made by doubling a 5-mm thick aluminum plate on part (2 m by 0.5 m) of M1 with adhesive agent and many rivets. M3 was an L-shaped structure that was made by cutting off part of M1.

The models of mooring devices are shown in Figure 7.13. These models were made from standing platforms in the water basin, representing the mooring dolphins, and were equipped with sliding devices with built-in linear springs. The connection between the VLFS model and the mooring-device model was made using a 50-cm-long rod with universal joints at both ends.

Table 7.4 Principal particulars of the model

	Prototype	1/100 model
Length	1,000 m	10 m
Breadth	200 m	2 m
Height	7 m	70 mm
Draft	1.5 m	15 mm
Displacement	307.5×10^3 t	300 kg
EI (Longitudinal)	1.46×10^{13} Nm^2	1.46×10^9 Nmm^2

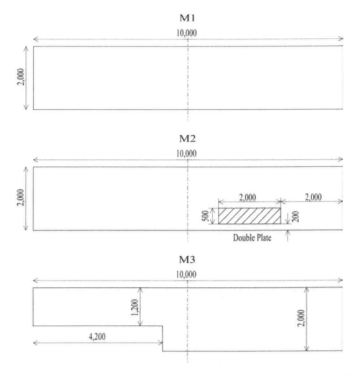

Figure 7.12 Plan views of M1, M2, and M3 models.

Figure 7.13 Photo of model mooring device.

Figure 7.14 Photo of Hechimaron.

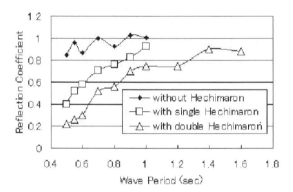

Figure 7.15 Wave reflection coefficients of breakwater model.

The breakwater model was made from 2.3-mm thick steel plates. In order to adjust the reflection coefficient of the breakwater, wave absorber called "Hechimaron" was fixed on both sides of the breakwater. A photograph of the wave absorber is shown in Figure 7.14. The measurement results of the reflection coefficient of the breakwater model are shown in Figure 7.15.

Experimental basin

The water basin used for the model experiments was the Ocean Engineering Basin of the National Maritime Research Institute. This water basin is

40 m long and 27 m wide and has a variable water depth of 0–2 m. It is equipped with a flap-type wave-making device on the short side of the basin. The depth of the water in which an actual VLFS is to be installed is assumed to be about 20 m. The corresponding water depth of the model scale is 20 cm deep, but it is difficult to generate waves by the flap-type wave maker in this water depth. Thus, the experiment was conducted in the 60-cm deep water. The corresponding elastic-response analysis was also carried out using the 60-cm water depth.

Experimental condition and measurement items

The measurement items and the measuring instrument units are listed in Table 7.5. The arrangement of sensors is shown in Figure 7.16. The conditions of the conducted experiments are shown in Figure 7.17.

The experiment conditions 1 through 3 are the basic conditions. The experimental conditions 4 through 12 correspond to the experiment purpose 1, that is, to study the mutual effect of a VLFS and a breakwater. The experimental conditions 14 through 18 correspond to the experiment

Table 7.5 Measuring items and measuring devices

Measuring items	Measuring devices	Measuring points
Vertical displacement	Potentiometer	102
Structural strain	Strain gauge	16
Horizontal displacement of mooring point	Laser-type displacement sensor	7
Wave height (incident)	Servo-type wave-height gauge	1
Wave height (around model)	Servo-type wave-height gauge	2
	Total	128

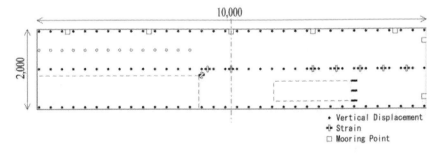

Figure 7.16 Plan view of sensors arrangement.

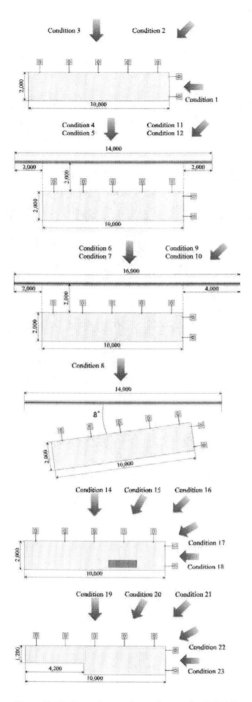

Figure 7.17 Experimental conditions for M1, M2, and M3 models.

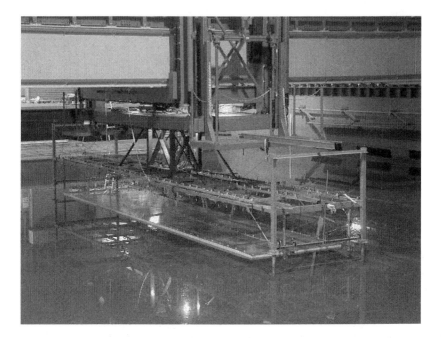

Figure 7.18 Photo of model experiment setup.

purpose 2, that is, to study rectangular VLFS of nonuniform rigidity. The experimental conditions 19 through 23 correspond to the experiment purpose 3, that is, to study nonrectangular VLFS of uniform rigidity. The periods of regular waves generated during the experiments were in the range of 0.5–1.4 s. The photographs of the setup conditions of the experimental models are shown in Figure 7.18.

Example of measured results

The elastic-response analysis results and the model experiment results agreed well. This confirms that the analysis method used was appropriate. Figure 7.19 shows the case of a 45° incident wave angle to the M3 model.

7.5.2   Elastic-response experiment for dynamic load

A VLFS undergoes elastic response not only to waves but also to dynamic loads such as the takeoff and landing of aircrafts. As an example of the model experiments of the elastic response of VLFS to dynamic loads, experiments of a pontoon-type VLFS subjected to impulsive load and moving load are introduced (Endo 1999).

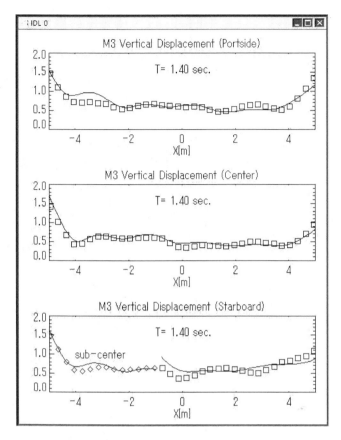

Figure 7.19 Vertical-displacement distribution measured (marks) and calculated (solid lines) in a case of 45° incident wave angle for model M3.

Purpose of experiment

In order to estimate VLFS' elastic response to the takeoff and landing loads of aircrafts, the analysis method for solving the elastic-response problems in time domain was developed (Ohmatsu 1998; Endo 2000). The purpose of the experiment is to verify the validity of these analysis methods.

Model

The VLFS for the experiment was the 300-m long floating structure used for the Mega-Float Project Phase 1 that was implemented by the Technological Research Association of Mega-Float. The scale of the model was 1/30.77 and the length of the model was 9.75 m. The complete dimensions of the

Table 7.6 Principal particulars of the model

	Prototype	1/30.77 model
Length	300 m	9.75 m
Breadth	60 m	1.9 m
Height	2 m	54.5 mm
Draft	0.5 m	16.6 mm
Displacement	9,225 t	307.5 kg
El (Longitudinal)	4.83×10^{11} Nm^2	1.75×10^{10} Nmm^2

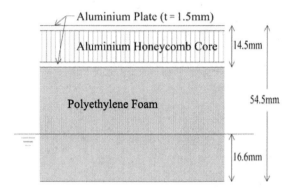

Figure 7.20 Cross section of model used for dynamic load experiments.

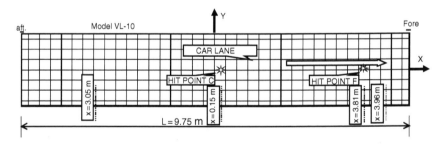

Figure 7.21 Plan view of model used for dynamic load experiments.

model are listed in Table 7.6. The cross section and the plan of the model are shown in Figures 7.20 and 7.21, respectively. The car lane shown in Figure 7.21 was set for a moving-load test and a weight was towed with a constant speed on the lane.

Experimental basin

The experimental basin used was the same as the previous one. But the experiment was conducted under the condition of 1.9-m water depth.

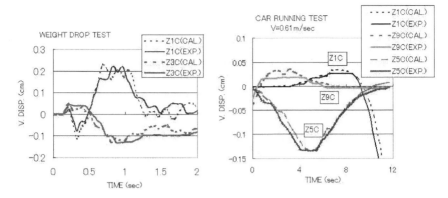

Figure 7.22 Time series of vertical displacement measured (solid lines) and calculated (broken lines) in weight drop test and car running test.

Experimental conditions and measured quantities

Two kinds of experiments were conducted. The first kind is the impulsive load experiment that involves dropping a weight on "Hit Point." The second kind is the moving load experiment that requires the towing of a weight on "Car Lane" with a constant speed. The dropped weight was 20 kg and was equivalent to 580 t in the actual VLFS. The towed load was 6.9 kg and the maximum towing speed was 0.61 cm/s which was equivalent to 12.2 km/h for the actual VLFS. The measured quantities were the acceleration of the weights and vertical displacement at 9 points.

Example of measured results

The results of the elastic-response analysis and the model experiment results agreed well. Therefore the appropriateness of the analysis method was validated. The example of the results of the impulsive experiment and the towing experiment are shown in Figure 7.22. In this figure, the results were computed by Endo (2000).

7.6 Concluding remarks

The experimental method using a large-scale model in a large experimental basin is elaborated in this chapter. It can be said that the model experimental method of a several kilometer long VLFS has been already established. On smaller scale, Kagemoto (1999) and Murai (2004) conducted an interesting experiment using a very small water basin. The model in this experiment is also very small such as an OHP film and a thin "balsa" plate. Although it is difficult to measure accurately the displacement, such an experiment

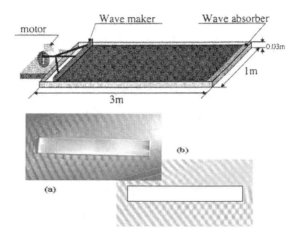

Figure 7.23 Very small water tank and wave pattern around small VLFS model. (a) Experiment, (b) Theoretical calculation.

Source: Murai (2004).

Figure 7.24 Wind tunnel test for estimation of wind load acting on a VLFS.

is suitable to learn the wave patterns around VLFS. Figure 7.23 shows an example of a very small water basin and the measured results.

As for the model experiments of VLFS, other than the elastic-response experiment in a water basin, there are wind tunnel experiments to study the effects of wind forces (Figure 7.24, Ohmatsu 1997) and hydraulic

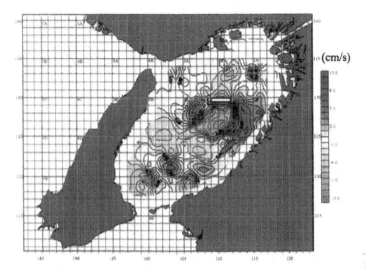

Figure 7.25   Residual current around VLFS measured by hydraulic experiment.
Source: Ueshima et al. (1998).

experiments to investigate ocean environment changes, such as current changes and current dispersion by VLFS (Figure 7.25, Ueshima 1998). The details of these experiments are shown in the references.

It is extremely important to investigate various phenomena by conducting experiments using a model that well simulates actual phenomena for the development of new technologies in order to realize the construction of actual VLFS.

References

Endo, H. (1991) "The laws of similitude in hydroelasticity problems," *Journal of the Society of Naval Architects of Japan*, 169, pp. 347–354 (in Japanese).

Endo, H. (2000) "The behavior of a VLFS and an airplane during takeoff/landing run in wave condition," *Marine Structures*, 13, pp. 477–491.

Endo, H. and Yago, K. (1999) "Time history response of a large floating structure subjected to dynamic load," *Journal of the Society of Naval Architects of Japan*, 186, pp. 369–376 (in Japanese).

Kagemoto, H., Murai, M., Fujino, M., Kato, T., and Kondo, Y. (1999) "Experiments of a very large floating structure in a very small water tank," *Proceedings of the Third International Workshop on Very Large Floating Structures*, Hawaii, Honolulu, pp. 555–561.

Murai, M. (2004) "Experiments on hydroelastic responses of a very large floating structure in a very small water tank," *Conference Proceedings of the Society of Naval Architects of Japan*, 3, pp. 131–132 (in Japanese).

Ohkawa, Y. (2000) "Various Mega-Float model experiments on elastic behavior in waves," *74th General Meeting of Ship Research Institute*, Mitaka, Tokyo, Japan, pp. 291–294 (in Japanese).

Ohmatsu, S. (1998) "Numerical calculation of hydroelastic behavior of VLFS in time domain," *Proceedings of the Second International Conference on Hydroelasticity in Marine Technology*, Fukuoka, Japan, pp. 89–97.

Ohmatsu, S. (2005) "Overview: research on wave loading and responses of VLFS," *Marine Structures*, 18, pp. 149–168.

Ohmatsu, S., Takai, R., and Sato, H. (1997) "On the wind and current forces acting on a very large floating structure," *Journal of Offshore Mechanics and Arctic Engineering*, 119(1), pp. 8–13.

Society of Naval Architects of Japan (1997) "Mega-Float offshore structure," Chapter 9, *Experimental Technique*, Seizando, pp. 303–342 (in Japanese).

Ueshima, H., Takarada, M., Hikai, A., Kobayashi, E., and Minamiura, S. (1998) "An experimental study on environmental impact of huge floating marine structure," *Proceedings of the 14th Ocean Engineering Symposium*, Tokyo, pp. 307–312 (in Japanese).

Yago, K. and Endo, H. (1996) "On the hydroelastic response of box-shaped floating structure with shallow draft (tank test with large scale model)," *Journal of the Society of Naval Architects of Japan*, 180, pp. 341–352 (in Japanese).

Chapter 8

Maintenance and anti-corrosion systems

*Eiichi Watanabe, Hitoshi Furuta, Makio Kayano,
Motohiko Nishibayashi, Tomoaki Utsunomiya,
Kunitomo Sugiura, and Masahiro Yamamoto*

8.1 Introduction

The maintenance of infrastructure has become urgent world-wide concern. Even for bridges alone, it is considered to be very hard economically, socially and technically to maintain all of them in a healthy condition because of the large size. One cannot optimistically expect sufficient budgets for the maintenance in view of the current unfavorable economic condition of infrastructure holders and people's indifference to the needs of such maintenance. Under these circumstances, nevertheless, the Asset Management System (AMS) is drawing much attention. The strategy for maintenance of offshore structures including very large floating structures (VLFS) should follow the similar principle as that for bridge and land-based structures. As compared with the land-based structures, offshore structures are considered to be much more vulnerable to corrosion. However, in the area of the VLFS, the maintenance policy has not been well established because of its relatively new history and the fact that only a few VLFSs are in existence. Thus, it is proposed that the necessary concepts of the AMS be adopted for VLFSs.

8.2 Long-term deterioration and reliability of structures

During their service lives, structures tend to deteriorate, become obsolete or functionally outofdate. To prevent or delay the aging of infrastructures, structural maintenance technology plays a very important role. The materials used in infrastructures break down without regular maintenance effort. They are destined to become old and degrade just like human bodies and thus maintenance is needed in order to prolongate their residual lives.

Sakai (1997) promoted the performance-based design concept, where durability and restorability are calling for much attention. The life-cycle cost may become one of the key factors. The processes of planning, design, construction, and maintenance are interrelated and they should be carried out interactively (Corotis et al. 2001). Yet there is a long way to go before the performance-based design is truly and effectively adopted.

Structures are planned and designed for the expected maximum loading during the whole service life. Natural load effects such as the winds, water waves, tsunamis, and earthquakes become critical in many cases. These dominating loads are described by means of the Return Period, T. The expected service life, Q, is used in the estimation of the probability of nonexceedance P_N that is defined as

$$P_N = \left(1 - \frac{1}{T}\right)^Q \tag{8.1}$$

Figure 8.1 illustrates the value of P_N as a function of Q/T, the ratio of the design service life and the return period in the case of $T = 50$ years. In many cases, Q/T, may be conventionally assumed to be 1/2 or 1/3. In such cases, the probability of nonexceedance, that is the probability of the designed load never occurring during the service life is estimated to be 61% or 70%, respectively from this figure. Thus, the recently adopted concept of the return period to be twice that of the design service life seems to be rather reasonable.

For the reliability of structures, another aspect, namely, the probabilistic characteristics of the strength and external loads should be considered. According to Frangopol and Parag (1999) and Thoft-Christensen (1999), the value of the reliability index β of a newly built structure varies between 5 and 12. Denoting the margin of safety by y, the standardized normal distribution $\Phi(y)$ is assumed to be given by

$$\Phi(y) = \int_{-\infty}^{y} \frac{1}{\sqrt{2\pi}} \exp\left\{-\frac{1}{2}(x - \bar{x})^2\right\} dx \tag{8.2}$$

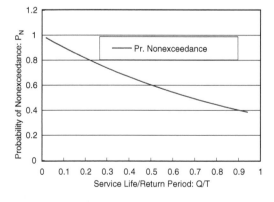

Figure 8.1 Probability of nonexceedence, P_N, depending on Q/T.

where x is the standardized variable and its expected value is $\bar{x}$. The probability of failure may be expressed by

$$\Phi(-\beta) = 1 - \Phi(\beta) \qquad (8.3)$$

For example, $\beta = 5$ corresponds to the probability of failure of $\Phi(-5) = 0.0000003$ while $\beta = 12$ corresponds to the probability failure of $\Phi(-12) = 0$. Structures will gradually lose initial soundness because of increasing cracks and corrosion after many years of service, resulting in a smaller value of β. A value of $\beta = 4.6$ may be regarded as a candidate for the lowest allowable limit. This value corresponds to the probability of failure $\Phi(-4.6) = 0.000002$ as shown by Figures 8.2 and 8.3. Below this value of β, essential maintenance is required. Figure 8.3 shows the basic degrading property of structures proposed by Frangopol et al. (2001).

After some time period t_1 from the completion of a structure, the deterioration starts to take place at the rate α. Even in the case where essential maintenance is not the immediate concern, preventive maintenance may have to be executed. Needless to say, his simple scheme of deterioration is

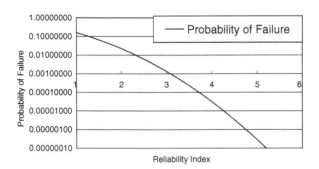

Figure 8.2   Probability of failure and reliability index, β.

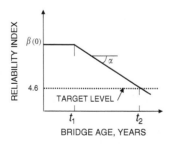

Figure 8.3   Reliability index, β and aging.

Source: Frangopol (1999), Thoft-Christensen (1999).

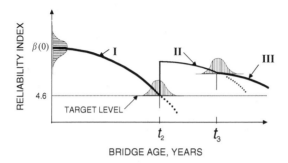

Figure 8.4 Conceptual degradation of structures.

Source: Frangopol and Parag (1999).

only a crude but simple linear approximation of more realistic phenomenon of certain nonlinearity and stochasticity as elucidated by Figure 8.4. As a matter of fact, there will also be random variations for the reliability index β and the time reaching the critical stage of the maintenance, t_2, depending on each structure.

In Figure 8.4, the initial path I ends at the time $t = t_2$. Then the reliability index β can be upgraded just after $t = t_2$ as if the structure were newly built and the deterioration process following along curve II. If a preventive maintenance schedule is executed at $t = t_3$, the rehabilitated structure will follow the path indicated by curve III with the reduced rate of the deterioration. Furthermore, it was indicated by Frangopol et al. (2001) that the reliability index β can drop down to between 2 and 3 after the service life of 40 years, corresponding to the reduced probability of failure P_f reaching the value of 0.023 and 0.0014, respectively.

8.3 Background, scope, and concepts in maintenance of structures

Structures deteriorate due to weather conditions and external loads. Water tends to corrode steel and can scour away bridge foundations. Meanwhile every car and truck that passes over a bridge causes it to deflect. Excessive loads can cause cracks that destroy the structural integrity. The bridge superstructure is susceptible to corrosion, water damage, and metal fatigue under repeated application of live loads. Despite the fact that engineers agree on the general mechanisms of structural failure, the details are not well understood. Bridge structures are too complex for complete computer analysis, and consequently, simulations require a good number of simplifying assumptions. This results in seldom matching the behavior in the field (Dunker and Rabbat 1993).

Once a structure has begun to deteriorate, the process of decay accelerates and is difficult to stop. The portions of metal girders that are under a larger stress corrode more rapidly, and stress concentrations increase as the thickness of sound metal decreases. Similarly, damaged structural members with reduced load-bearing capacity become more vulnerable to the effects of heavy loading. Furthermore, debris-clogged joints prevent the movements necessary to relieve stresses of girders, and the speed, surface roughness, and suspension of vehicles interact to amplify stresses on a bridge (Dunker and Rabbat 1993).

Draining the water always poses serious problems. Water has to be removed from structural elements and bearing surfaces, where the combination of salt and stress can lead to destabilization of the structures in a few years. Moreover, debris inhibits deck drainage; water and deicers corrode steel reinforcing bars, causing spalling, where standing water on the slab accelerates the deck deterioration and the improper drainage causes damage to concrete. Complex drainage systems, however, are expensive and require periodic maintenance to keep them from clogging.

In the maintenance of structures, various aspects are considered. These aspects should desirably be considered at the very beginning of the new construction stage. In some cases, several specific items are much more important than the others. The keywords include such items as listed in Table 8.1.

Furthermore, not only the mechanical structural conditions, such as whether there exist corrosions and fatigue cracks but also whether the value of intended functions of the structure matches with the needs of the present and future society should be considered. Sometimes the historical value may be evaluated leading to the historical retrofitting of structures.

Table 8.1 Keywords: background and scope in maintenance of structures

	Keywords
Ideal	Aesthetics, comfortability, constructionability, cultural heritage, durability, ecology, economy historic structures, life-span extension, long-term reliability, maintainability, environmental friendliness, redundancy, restorability, safety, serviceability, sustainability, vulnerability
Avoid	Aging, cracks, fatigue
Toward	Easy demolition, high performance, intelligent structures, life cycle, performance-based structural engineering, short construction time
Method	Composite materials

8.4   Maintenance procedures

8.4.1   Structural monitoring, assessment, and site investigation

The first action by the owners of structures would be monitoring or inspections for defects. Their results are described in reports, which are diagnosed and evaluated afterwards. Then treatment, repair, rehabilitation or complete reconstruction follow. This flow, nevertheless, has many difficulties. First of all, in many cases, neither the inspectors nor the evaluators may be well familiar with the expertise. In this regard, training such as periodic holding of educational seminars, workshops and exhibitions on the advanced reconnaissance technologies may be beneficial and worthwhile.

Monitoring technologies should be encouraged. What kind of items and spots should be inspected and how? Sometimes getting access to the inspection spots can prove difficult. The modern trend would be "easy access," "easy monitoring," and "easy data acquisition" making use of advanced technologies with the application of the advanced micro-devices or sensors and information technologies.

Figure 8.5 elucidates the maintenance problems and their countermeasure. Furthermore, the construction of the database in combination with the inventories of structures may be highly recommended for systematic long-term maintenance. The inspection and the monitoring regimen provide essential information on the health conditions of structures. With the advent of advanced sensors and information technologies, data acquisition on the structures has become much easier and accurate. Moreover, with the advances in fracture mechanics, diagnosis technology has greatly improved. Thus, it is now possible to predict residual structural life. Table 8.2 shows the keywords associated with structural monitoring and health assessments of structures.

8.4.2   Structural repair, rehabilitation, strengthening, and reconstruction

Based on the previous procedure for monitoring and health assessment, the enforcement of repair, rehabilitation, strengthening, retrofitting or reconstruction of structures will follow. Table 8.3 lists the keywords in this process. As the cost for this process is soaring, it should be considered at the time of planning and design. Several technologies have become quite popular such as the use of CFRP or polymers, rubber bearings, and seismic retrofitting.

Obviously how to maintain and preserve existing infrastructure is highly important for the benefit of people. In spite of frequently published versions of bridge design specifications (Japan Road Association 2002), no

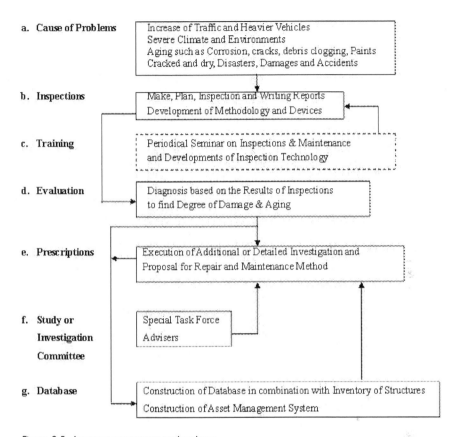

a. **Cause of Problems**

Increase of Traffic and Heavier Vehicles
Severe Climate and Environments
Aging such as Corrosion, cracks, debris clogging, Paints
Cracked and dry, Disasters, Damages and Accidents

b. **Inspections**

Make, Plan, Inspection and Writing Reports
Development of Methodology and Devices

c. **Training**

Periodical Seminar on Inspections & Maintenance
and Developments of Inspection Technology

d. **Evaluation**

Diagnosis based on the Results of Inspections
to find Degree of Damage & Aging

e. **Prescriptions**

Execution of Additional or Detailed Investigation and
Proposal for Repair and Maintenance Method

f. **Study or Investigation Committee**

Special Task Force
Advisers

g. **Database**

Construction of Database in combination with Inventory of Structures
Construction of Asset Management System

Figure 8.5 Asset management technology.

specifications have been enacted for existing structures (Urban Expressway 2004). Thus great efforts have been made in many places but things are not so easy. Even after the accident of the Silver Bridge in 1967 and subsequent accidents, the budget was not necessarily increased quickly. One had to wait until 1980s when the gasoline taxation started to be increased gradually in order for funds to be allotted to necessary road maintenance funds. The US Congress passed the Federal-Aid Highway Act (FHWA) and the National Bridge Inspection Standards (NBIS) was initiated for the training of bridge inspectors (Dunker and Rabbat 1993; Takagi 2000; Frangopol et al. 2001). Likewise, in Korea a similar trend for budgeting has occurred since the collapse of the Seongsu Bridge in 1994 (King et al. 1999). European countries are also making much effort in this area.

The predominant bridge management system in US is PONTIS, which was developed through a collaborative work between the FHWA and the State

Table 8.2 Keywords for structural monitoring, assessment and site investigation

	Keywords
Avoid	Alkalinization, carbonation, chloride, corrosion, cracking, deficiencies, deterioration, scour
Effects	Live-load effects, long-term deflections, truck-load effects
Technology	Acoustic emission, acoustic monitoring system, ambient-vibration technique, automatic monitoring, bridge-rating expert system, computer-aided information-expert system, damage detection, damage estimation, damage assessment, detection, diagnosis, duct injection, elastometric bearings, embedded micro-devices, fiber optic sensors, field survey or study, field testing, fracture critical members, fracture mechanics, global position system (GPS), health monitoring, information technology, inspection, instrumentation, internet and intranet, life-time prediction, long-term monitoring, micro-electro-mechanical systems, MEMS, monitoring, non-destructive tests, optical fiber sensors, passive peak sensor, portable instrumentation, probabilistic-based assessment, remote monitoring, quality assessment, rating, sensor, smart aggregate devices, stress range, structural health monitoring, system identification, ultrasound, X-ray

Table 8.3 Keywords for repair, rehabilitations, strengthening and reconstruction of structures

	Keywords
Scope	Addition, enlargement, improvements, modernization, noise reduction, reconstruction, refurbishments, recycling, rehabilitation, repair, replacement, restoration, retrofitting, seismic retrofit, shear-strengthening, shielding/permeability, transformation, widening
Material	CFRP, FRP, multicoat painting, polymers
Methodology	Rigid connection of girder ends, corrosion inhibitors, damper control, drainage, dry-air injection, external prestressing, gousing, post-strengthening, post-tensioning, pre-existing cracks, reinforcement bonding, stop hole, smart structures, stiffeners, strengthening, tendons, tungsten inert gas dressing, welding
Target	Fracture critical members, old structures

Department of Transportation. Furthermore, BRIDGIT was developed also in US by the National Cooperative Highway Research Program (Frangopol and Parag 1999). The former has been adopted in most of Department of Transportation (DOT) based on life-cycle cost (LCC). These two systems are utilizing element-level inspections, predicting the future condition of the elements in the network based on Markov chains and optimizing long-term expenditures for the preservation and improvement of the highway network. Outside the US, management systems have been developed both in Denmark and Finland. Furthermore, a number of bridge management systems have been developed for the commercial markets (Frangopol et al. 2001).

However, the following problems exist in the maintenance:

1 Astronomically large budget required to take care of great volume of infrastructure.
2 Lack of experienced engineers and good workers.
3 Lack of dependable source data.
4 Technical problems, where neither the assessment of wholesomeness of existing facilities, the selection of the feasible repairing methods nor the prediction of the future deterioration is easy.
5 Institutional Problems, where unlike newly built structures, there are many difficult problems to be improved such as in order system and cost-estimation method.

One way to overcome these problems is to introduce the concept of LCC and to establish a related AMS making use of the knowledge for predicting the future degradation of structural durability by experienced engineers (Matsumura 2006). Figure 8.6 summarizes such an AMS. Currently, six items may be treated, that is (a) paint coating, (b) pavement, (c) expansion joints, (d) slab, (e) concrete structural members, and (f) steel structural members. The concept of LCC will be explained subsequently (Nishibayashi et al. 2006; Yamamoto et al. 2006).

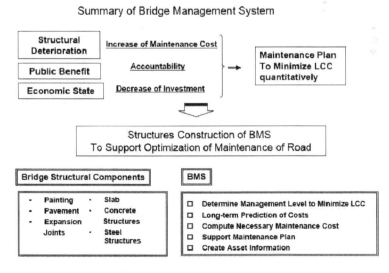

Figure 8.6 Summary of AMS.

It must be borne in mind that an AMS is a never-ending repetitive process such as shown in Figure 8.7. Furthermore, Figure 8.8 shows the construction of an AMS that is primarily composed of (a) management of basic information, (b) computation of LCC, (c) overall assessment of road structures, and (d) optimization of social load including the prioritization

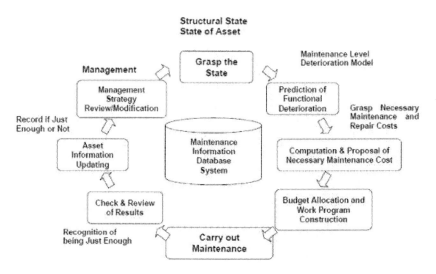

Figure 8.7 Asset management system.

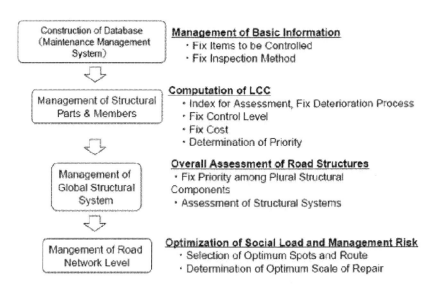

Figure 8.8 Construction of asset management system.

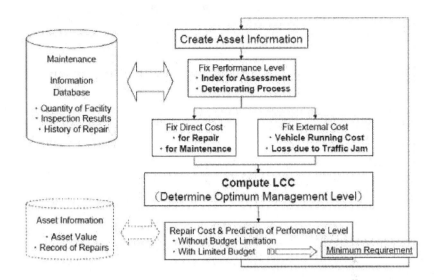

Figure 8.9 Structure of AMS.

and management risk. Consequently, use of the computer and database systems becomes a necessity as shown in Figure 8.9.

8.5 LCC

The LCC is referred to as the total cost over the life cycle. To be more specific, it refers to the total sum of the cost including planning, design, construction, maintenance, and replacement. The concept of LCC is not new and has been commonly used in the fields of electrical appliances, airplanes and machineries.

In the past, the concept of LCC for infrastructure has never been noted because of their large scale, long service life, and specialty of construction sites. Nowadays, however, the long-term vision for more economical and reasonable maintenance and management is being strongly felt to be of more importance because of recent shocking reports on the frequent deteriorations, damages, and the need for replacement.

Technically the prediction of future deterioration of the structures is required in order to assess LCC in addition to the assessment of the present state. Prediction of future society and the economy are necessary in spite of their uncertainties. Although it is perceived that a longer structural life and greater safety may result in a better quality of structure, in general better quality is assured with more expense. When the LCC is considered, there exists however a certain optimal performance level. Namely, when the performance is too low, the structure tends to deteriorate easily in spite of being economical at the initial stage. However, on the contrary, if it is too high, the structure becomes too expensive due to unduly frequent monitoring and inspection.

The LCC is usually computed by the following equation:

$$LCC = CI + CM + CF \times Pf \tag{8.4}$$

where CI is the initial construction cost; CM the maintenance and management cost; CF the failure cost and Pf the probability of failure. Figure 8.10 shows the change of the LCC as the performance level is changed and as the interval of maintenance is shortened, indicating that there may be a certain optimal point of the optimal LCC. In this figure, the external cost takes into account the vehicle running cost as well as the loss due to traffic jams.

Nevertheless, even in the evaluation of a direct LCC, the evaluation of Pf, CM and CF is not so easy because, first, Pf represents the probability for uncertain future damage or failure. Second, CM depends on a highly uncertain prediction of the deterioration and thirdly, the rigorous computation of CF is even more difficult. For example in the computation of CM alone, the assessment of effectiveness of the maintenance, damage evaluation of

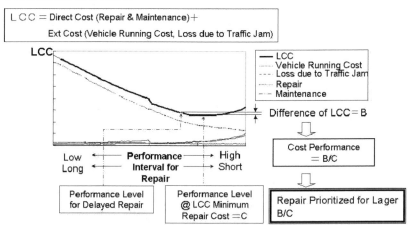

Figure 8.10 Change of LCC as the performance level is changed and the interval of maintenance is shortened.

Source: Nishibayashi et al. 2006, Division of Road Asset Management System 2003.

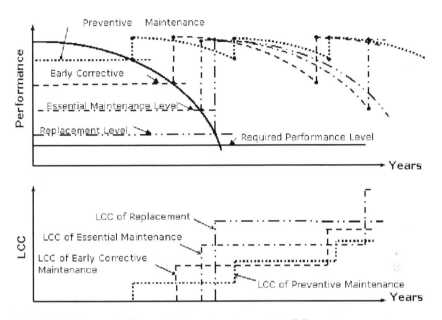

Figure 8.11 Various maintenance levels and corresponding LCC.

existing structures, prediction of future deterioration, definition of service life and its reasonable determination, evaluation of being economical, interest rates, and the prediction of economy are not simple. Figure 8.11 shows the relationship between the performance level and the LCC. For maintenance purpose, the early maintenance level, called the preventive maintenance level, may be prescribed at a relatively higher and more conservative level. But there may be such levels under which the structure may start to exhibit certain undesirable and unsafe behaviors. This level may be referred to as the essential maintenance level. In fact, the repair work below this level may require significantly larger amounts of labor and cost. If the deterioration drops below the replacement level, the repair becomes meaningless and thus the structure should be replaced.

8.6 Performance-based design and the LCC

An AMS is based on the LCC. The advantage of this system lies in that it leads to the quantitative evaluation of the maintenance cost during the whole service life. It must be admitted that great effort has been made already to reduce the cost for shorter periods and any further reduction of the cost implies the reduction of the quality too. For any further decisive reduction of the cost, a long-term vision is necessary. Lastly, for insisting on the needs for an AMS it may serve as the reliable background for the exhibition of information to citizens.

Currently, the performance-based design methods are drawing much attention world-wide (Sakai 1997; Floating Bridge Committee 2006). It is becoming a world-wide conviction that the maintenance problems are the most effectively treated in the performance-based design. The LCC surely serves as the measure to the provision of the target performance and the assessment of the satisfaction of the require performance.

8.7   Effect of restriction on budget for repair and long-term budgeting plan

In the case of no budget restrictions for repair, the cost will be determined by the optimal LCC and thus there will exist wide sporadic variations depending on each fiscal year. However, when the budget is restricted, such sporadic variations disappear but the characteristics will depend very sensitively on the amount of the budget. For example, if the budge limit is set at JPY 1.0 billion per year for pavement the repair must be undertaken which leads to the exhaustion of the whole budget as shown in Figure 8.12.

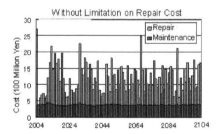

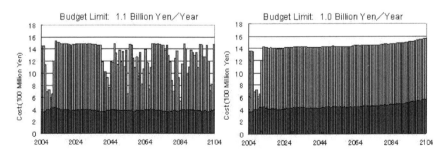

Figure 8.12 Pavement repair costs due to budget restriction.

Source: Courtesy of Hanshin Expressway Co. Ltd (Division of Road Asset Management System 2003).

If the repair is postponed due to budget prioritization, the performance of the pavement, Maintenance Control Index (MCI) will decrease. However, if the budget is appropriate, the MCI will be improved to maintain certain levels. If the budget is JPY 1.1 billion per year, the MCI will be properly maintained.

However, if the budget is either JPY 0.9 billion or JPY 1.0 billion per year, the MCI will apparently decrease from a certain time point. Figure 8.13 shows that the minimum MCI rapidly drops down to a zero value in 100 years when only JPY 0.9 billion per year is allotted for repair work. This implies that there is a real danger of loosing the performance when the budget for repair is reduced too much. Thus, it is recommended that the asset information monitored continuously for the long-term maintenance strategy and to prepare for additional allowances for the future repairs. If the net asset value is considered to be the total asset value subtracted by the allowance, the net asset value decreases with the deterioration of structures and recovered by the repair. In other words, if no repair is provided, the allowance will never be eliminated and the net asset value will not recover. The shortage of budget for the repair will become apparent as shown in Figure 8.14.

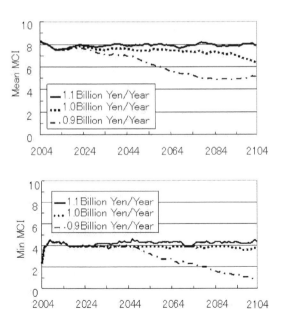

Figure 8.13 Change of performance (MCI) with budget for repair.

Source: Courtesy of Hanshin Expressway Co. Ltd (Division of Road Asset Management System 2003).

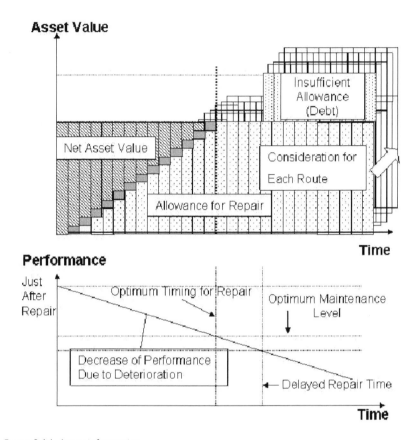

Figure 8.14 Asset information.

Source: Courtesy of Hanshin Expressway Co. Ltd (Division of Road Asset Management System 2003).

It is generally agreed that the LCC leads to the most economical budget-ing for an AMS. However there exists a difficulty for owners of structures. Although the long-term LCC is satisfactory, there may be some periods of time during which rather concentrated repair cost may have to be paid. Because of this, the maintenance scenario sometimes may have to be changed in order to cut the peak value of the LCC. Taking these into account, sev-eral improvements have been proposed. When the LCC is not uniformly distributed but concentrated in a particular period, the peak value of the LCC should be conveniently cut off so as to get a uniform budget plan (Kaneuji et al. 2006). The peak value can be reduced by changing the maintenance scenarios adopting multilevel budgeting. Figure 8.15 shows examples of such improvements. This improvement has been successfully

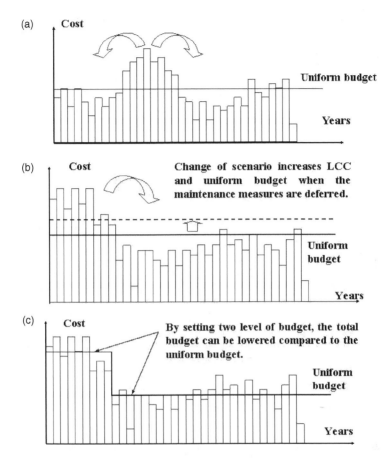

Figure 8.15 Application of multi-budgeting for cutting peak LCC. (a) Changing maintenance scenario for a uniform budget. (b) Change of scenario increases uniform
budget. (c) Budget increase reduced by means of multi-level budgeting.

Source: Kaneuji et al. 2006.

adopted in the construction of an AMS by the Aomori Prefectural
Government.

8.8   Items for maintenance peculiar to floating structures

So far basic principles for maintenance of land-based structures have been
described. In addition to these, there are many items specific to floating structures. Thus, the following inspections are thought to be specific to floating
structures (Floating Bridge Committee 2006).

8.8.1   Floating body

a   *Submergence*: Observation of the interior by naked eyes is easier but the use of sensors is recommended. In the case of submergence, water should be pumped out forcibly and immediately.

b   *Change of draft*: Even in the case of no submergence, the increased thickness of the slab, added facilities, and increased marine growth weight may result in larger draft.

c   *Corrosion protection*: In case of steel structures, the cathodic protection, heavy-duty paint coating, and metal coatings (for example, using stainless steel or titanium) are provided. For cathodic protection, the voltage is measured. Naked-eye inspection may be provided to find deterioration of paint and metal coatings. The eye inspection may also be provided to find the corrosion due to damage or spalling caused by the collision with littoral transports or floating bodies. Periodical inspection of submerged surfaces may be performed by divers.

d   *Marine growth on the surface*: Marine growth may be peeled off if they are accumulated due to their own weight. However, periodical removal and surface repair are desirable. During the removal, environmental consideration is necessary.

8.8.2   Mooring structure

a   *Reaction walls, dolphins, mooring piles*: Change of touching points with fenders and existence of cracks in the concrete part should be visually checked. Electric voltage may have to be measured to check if cathodic protection for the steel piles is working well. Inspection must also be provided to see if the surface of the reaction wall is worn out due to the motion of floating body caused by the tidal change. The clearance between fenders and dolphins must also be inspected to see if it is sufficient and that no obstacles or debris are trapped there.

b   *Fenders*: Usually fenders are located near the splash zone that suffers the most deterioration. Inspection must be provided to see if there is any corrosion, loosening or irregularity of metallic devices. For fender rubbers, the functionally critical depth of flaw should be found beforehand from the manufacturer. Their maintenance is very important because it is very expensive and may be subject to deterioration and replacement in a long period of time.

c   *Mooring rollers*: Naked-eye observation should be provided to see if they move smoothly without any noise or damage. The timing for replacement should be properly determined.

d   *Chains and anchor cables*: Chains are designed taking into account the reduction of the cross section due to corrosion. Inspection is performed with respect to the diameter of the cross-section to know that the corrosion speed is within the predicted value at different cross sections in the air, at the splash zone and underwater. Anchorage must be inspected if it has not moved after a high tide or a tsunami.

8.8.3 Connecting structures

a *Transient girders (connecting girders)*: The transient girders must be inspected to ensure that they have not undergone fracture, deformation and unusual noise as they are always subject to dynamic motion due to tidal change, motions by waves and winds.
b *Expansion joints*: Inspection is necessary to see if there are fractures, cracks, deformation, unusual noises, severe wearing at the sliding surface, and movement outside the allowable range. The timing, order, and the method for the replacement should be planned beforehand.
c *Bearing devices*: Inspection is necessary to see if there are fracture, cracks, deformation, unusual noises, severe wearing and hindrance to the proper function caused by the inclusion of debris inside the expansion board. The timing, order, and the method for the replacement should be planned beforehand.

8.8.4 Environmental change

The effects of the following environments on the floating bridge must be recorded to find the correct solution:

a water quality, stream velocity, and depth of water
b traffic volume

8.9 Inspection for damage caused under abnormal conditions

In addition to the inspections following abnormal conditions impacting on land-based structures such as earthquakes, typhoons, intensive rains, and heavy snow, the inspections of VLFSs must be performed after natural disasters such as tidal waves and tsunamis, and accidents caused by the collisions with ships and floating obstacles.

8.10 Paint coating strategy for offshore steel structures

In Japan, many steel structures are constructed near rivers and coasts. The speed of corrosion varies however from a freshwater zone to a saltwater zone. The dissolved oxygen in water, either fresh or salty, controls the corrosion of steel. In saltwater, which has a higher electric conductivity than freshwater, the different amount of oxygen supply between the zone above water level and in the water immediately below causes the corrosion due to the formation of macro cells. Furthermore, the generated rust tends to accelerate the corrosion of steel members by the repetitive process of oxidization and deoxidization. Lastly, the waves and littoral transport in the sea cause the formed rust to peel off and thus accelerates the corrosion faster than in a freshwater environment.

Although in recent years, corrosion protection technology has improved significantly in Japan, the problem of corrosion is a long-term process and thus it might take a long time before the effects of the corrosion protection methods are assessed. Consequently, the validity of the short-term prediction methods is among the top research targets. In the meantime, asset management is becoming increasingly important and the optimum methods for the least LCC are discussed in many fields (JASBC 2001; JSSC 2002).

8.10.1   Basics of corrosion

Ionization is the basis for understanding corrosion. This refers to the ordering of metals arranged in order of susceptibiliy to ionization or oxidization. The most susceptible metal is K (potassium)and is positioned to the left end and the most noble metal is Au (gold) on the right end.

Base metal $\leftarrow$ $\hspace{10em}$ $\rightarrow$ Noble metal

$$K < Na < Mg < Al < Zn < Cr < Fe < Ni < Sn < (H) < Cu < Ag < Pt < Au$$

When two pieces of metals, namely, Metal A and Metal B (assuming Metal B is nobler than Metal A) are put into the electrolyte and connected to each other by a wire, the less nobler Metal A tends to be ionized after losing the electrons and dissolving in the electrolyte as M^+ as shown in Figure 8.16 and the electric current starts to flow from right to left. Conversely, the minus ion moves through the wire from left to right instead. This phenomenon is referred to as contact corrosion.

8.10.2   Cathodic protection galvanic anode

It may be quite natural to think that by reversing the above-mentioned process, the corrosion may be stopped. Figure 8.17 shows the basics of the cathodic protection method. In this figure, metal Al is used for the galvanic anode.

8.10.3   Corrosion in ocean environment

Seawater contains about 3.5% of salt. Its electric conductivity is high and oxygen is always supplied from the air. Although seawater's pH is roughly 8

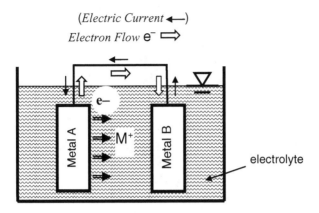

Figure 8.16 Basics of corrosion.

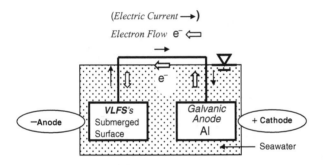

Figure 8.17 Basics of cathodic protection.

(rather neutral), the corrosion rate of steel at the splash zone or just beneath the ebb and flow zone is larger than in the fresh water (JASBC 2001). The rate of corrosion along the depth of water depends on the environment. Table 8.4 gives the standard values of the rate of corrosion and Figure 8.18 shows a sketch of the distribution of corrosion along the depth of water and ground (Abe 2001; JASBC 2001).

The following may assist in the understanding the corrosion by seawater (JASBC 2002; Kozai Club 2001):

1 The surface is easy to get wet because of the deliquescence of adhered salt.
2 Frequent repetition of dry and wet conditions due to splashing.
3 Waves always stir seawater and intensify the dissolved oxygen and help forming the macro cell as shown in Figures 8.18 and 8.19.

Table 8.4 Rate of corrosion

Location	Corrosive environment	Corrosion rate (mm/year)
Offshore side	Above HWL	0.3
	HWL to −1 m below LWL	from 0.1 to 0.3
	−1 m below LWL to seabed	from 0.1 to 0.2
	in mud layer beneath seabed	0.03
Onshore side	in air	0.1
	in earth above water level	0.03
	in earth below water level	0.02

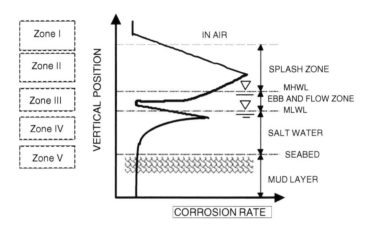

Figure 8.18 Distribution of corrosion rate of steel.

In Zone I: Owing to the deoxidization of dissolved oxygen and large supply of dissolved oxygen, the following chemical reaction takes place: $\frac{1}{2}O_2 + H_2O + 2e^- \rightarrow 2OH^-$

In Zone II: Corrosion of iron takes place such that $Fe \rightarrow Fe^{2+} \leftrightarrow Fe^{3+}$

In Zone III: The macro cathode is formed because of large supply of dissolved oxygen.

In Zone IV: The macro anode is formed as there is only little supply of dissolved oxygen.

The marked local corrosions in the sea environment occur in either the splash zone above the Mean High Water Level (MHWL) or the zone immediately below the Mean Low Water Level (MLWL). In the former zone, the rusted layer and the attached marine organisms accelerate the corrosion while in the latter zone, the oxygen cell is formed between the ebb and flow zone. Furthermore, the process of corrosion accelerates in the former zone

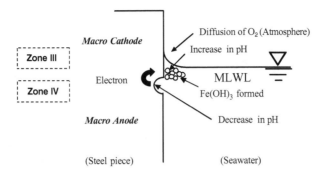

Figure 8.19 Localized corrosion process for ordinary carbon steel in seawater environment.

due to the temperature increase as a result of sunlight radiation, the adherence and concentration of salt at the surface, the repetition of dry and wet conditions, the rich supply of oxygen from the air, and the peeling-off of marine growths by strong waves. Generally speaking, the environment in the earth layer below the seabed is even milder, although it depends on the salt density, the degree of contamination, and the meteorological condition. As shown in Figure 8.19, the ebb and flow zone becomes the cathode due to a faster diffusion of oxygen and the adjacent zone immediately below the MLWL becomes the anode. Accumulated rust in the anode hinders the supply of oxygen and thus the corrosion is accelerated due to the macro cell.

What about the case of floating structures instead of fixed offshore structures? Since the ebb and flow zone does not exist because of the constant draft of floating structures, the distribution of corrosion may be different from that for fixed offshore structure as shown in Figure 8.18. Unfortunately, there has been hitherto no such information, suggesting the need for conducting exposure tests for floating structures.

8.10.4 Strategy for corrosion protection

Because of the severe corrosive environment, corrosion protection should be made, especially for the parts just below the MLWL when severe local corrosion occurs. For such parts, cathodic protection is generally applied, whereas the coating methods are applied for parts shallower than 1 m below the MLWL. The coating methods include (1) organic lining; (2) inorganic lining; (3) petrolatum lining; (4) paint coating; and (5) cathodic protection as described earlier. These methods may be tabulated in Table 8.5. The inorganic linings include metal linings such as titanium-clad lining (JTS 2000; Nippon Steel 2000), stainless steel lining (PWRI and JISF 2006;

Table 8.5 Corrosion protection methods

Classification	Method	Comments
Organic lining	a. Underwater hardening lining b. Special urethane polymer lining c. Ultra-thick mastic d. Polyethylene lining	Relatively thicker coating of 2–10 mm thick
Inorganic lining	a. Metal lining b. Mortar lining with titanium cover c. Crus (underwater stud welding & RC) d. Electrocoating e. Ductal* cover lining with high-strength fiber f. Metal injection	Mechanically stronger; better anti-shock; anti-wearing; better glossy look; *Ductal is a fiber reinforced hybrid material using reactive powders
Petrolatum lining	a. Petrolatum protection b. Monolithic foaming c. Titanium plating d. Stainless-steel cover	Easier underwater work; easier treatment of base layer; curing is not necessary
Paint coating	a. Lining with glued sheet b. Underwater-hardening epoxy resin c. Ultra-thick epoxy resin d. Tar-epoxy resin e. Glass-flake epoxy resin f. Glass-flake polyester resin g. Epoxy-resin paint coating for wet surface h. Glass-flake polyester resin/fluorine resin i. Water-resolving thick painting/fluorine resin j. Thick-coating inorganic zinc/epoxy/polyurethane resin	Thicker than for land-based structures
Cathodic protection	a. Galvanic anode or sacrificial anode (aluminum alloy anode) b. Method by external power source	Highly reliable; effective for underwater and underground; economical; good for newly built and existing structures; life of protection system can be freely selected; however for large-scale structures the method with external source is better

Source: Courtesy of PARI, CDIT & JASPP.

Table 8.6 Results of exposure tests at offshore experimental jacket

Zone	Ordinary stainless steel SUS304 & SUS316 or their equivalent stainless steel	High-performance stainless steel satisfying: Cr+3Mo+10N≥38 (mass%)
Splash zone	Local corrosion found	
Ebb and flow zone Underwater	No corrosion found due to simultaneous cathodic protection	No local corrosion found

Source: PWRI and JISF (2006).

Sato et al. 2003), and thermal spraying using zinc, aluminum and aluminum alloys (JAPH 1999).

The titanium-clad lining has been successfully adopted for the steel bridge piers of Tokyo Bay Crossing Road (nicknamed Aqualine) and for the steel pontoons of Yumemai Bridge in Osaka (Watanabe 2003; Watanabe et al. 2001). While the stainless-steel lining has been also successfully adopted at Ohi Container Terminal based on exposure tests conducted for 20 years at a experimental station at the offshore of Ohi River in Shizuoka. Table 8.6 shows the results of two different stainless steels. It was found that the seawater-resistant stainless steel containing Cr, Mo, and N did not show any local corrosion; however, ordinary stainless steel showed local corrosion at the splash zone (PWRI and JISF 2006; Sato et al. 2003). Based on this finding the heavy-duty stainless steel is going to be used for the columns of steel jackets of a new runway at Haneda Airport.

Table 8.7 compares the feasibility of various corrosion protection methods for steel sheet piles in coastal area (http://www.jaspp.com/koyaita/qa23.html) by courtesy of the Japanese Association of Steel Piles.

8.10.5   Historical developments of corrosion protection

Table 8.8 shows the chronological developments of corrosion protection methods that were mainly developed by CDIT et al. between the 1960s and the 2000s.

8.10.6   Performance of various paint coatings

Exposure test

Although corrosion prevention for steel structures is very important in coastal areas, few studies have been reported until recently (Abe 2001; Masubuchi 1999; Muranaka 1998; Watanabe 2006; Yamamoto et al. 1986). Based on a 23-year corrosion exposure test on a 9-m long rolled

Table 8.7 Feasibility of corrosion protection methods for harbor facilities (steel sheet piles) depending on environment

	In air		Near water surface		Underwater		Under seabed		Underground at onshore side	
	A	B	A	B	A	B	A	B	A	B
Paint coating	5	5	4	3	4	2	4	2	4	2
Lining	5	5	5	4	4	2	4	2	4	2
Concrete coating	5	5	5	4	3	2	3	2	3	2
Lining with anti-corrosion metal	5	5	5	3	4	2	3	2	3	2
Cathodic protection	2	2	4	4	5	5	5	5	5	5

Source: Courtesy of Japanese Association of Steel Pipe Piles (http://www.jaspp.com/koyaita/qa23.html).

Notes
A: Newly built structures; B: Existing structures.
5: Work possible; effective; 4: Work possible; fairly effective; 3: Work possible; not so effective or economically not recommendable; 2: Work impossible; not effective.

Table 8.8 Historical developments of corrosion protection

Years	Types	Developments	Remarks
1960s	Linings	Tar-epoxy-resin paint coating zinc-rich paint	a. Protection by oil-based paint
	Cathodic protection	a. External power source (early 1960s) b. Galvanic anode	b. Shift from external power source to galvanic anode c. No good results for either anti-seawater or weathering steel
	Others	a. Anti-seawater steel (US steel) b. Weathering steel	d. Primarily concept of using additional corrosion margin e. No. of instances of intensive corrosions started to increase
1970s	Linings	a. Rubber chloride b. Urethane resin	a. Cathodic protection started to be specified in design rule to newly built structures 1976
	Cathodic protection	a. Superior aluminum anode developed: (2000–2300 Ah/kg) → (2600–2700 Ah/kg)	b. Improvement in underwater welding for shortening construction period and securing safety and installing galvanic anodes
	Others	Popularization of galvanic anode	
1980s	Linings	a. Cement mortar/FRP cover b. Polyethylene/polyurethane cover; underwater-hardening lining; petrolatum lining; heavy-duty coating; ultra-thick epoxy resin	a. Accidents at kurihama pier No. 2 and yamashita pier (1981) b. Survey conducted all over Japan by Ministry of Port and Harbor: CDIT founded (1983)

Table 8.8 Continued

Years	Types	Developments	Remarks
	Cathodic protection Others	c. Fluorine-paint coating developed d. Electro-coating developed Examples of corroded pipe piles published	c. Corrosion protection and repair manual for port and harbor steel structures (1986) d. Revision of technical standard & commentary for port & harbor facilities (1989) Application of CP to existing structures; various protections using linings developed
1990s	Linings	a. Titanium applied for lining cover b. Titanium-clad steel applied c. Application of electro coating d. Monitoring tests initiated for impedance and insulation resistance	a. Revision of corrosion protection & repair manual for port & harbor steel structures (1997) b. Revision of technical standard & commentary for port & harbor facilities (1999)
	Cathodic protection Others	a. Maintenance & repair manual for port & harbor structures b. Corrosion protection & repair manual for port & harbor steel structures c. Study on effectiveness of CP against side erosion* d. Study on relationship between significant wave height and cathodic current and voltage	c. Maintenance & repair manual for port & harbor structures (1999) Titanium and electro coating started to be used for corrosion protection
2000s	Linings	Practical use of stainless-steel lining	a. Tests of petrolatum lining using anti-corrosional stainless steel started b. Hybrid materials of high performance cement and steel fibers started c. Exposure test using metal injection
	Cathodic protection Others	Examples metal linings published	CP test conducted on pipe piles with stainless steel lining Detailed study after 20 years of exposure tests/report published in 2004

Source: Courtesy of CDIT (Coastal Development Institute of Technology) (PARI, CDIT & JASPP, June 6, 2006).

Note

*Side erosion refers to the phenomenon of acceleration of corrosion due to rust peeling off by sand.

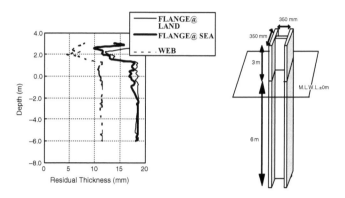

Figure 8.20 Variation of thickness in vertical direction.

H member made of ordinary carbon steel (the nominal yielding strength of 240 MPa) with the cross section of 350 mm × 350 mm, flange thickness of 19 mm and web thickness of 12 mm and the wetted depth of 6 m (cross-sectional area = 17,500 mm^2) (Yamamoto 1991 and 1992), a study was conducted to find the strategy for an effective maintenance of a wharf which is constructed from steel rolled-H sections. Figure 8.20 shows some results of the exposure tests. It may be seen that the decrease of thickness of steel reached 8–9 mm after 23 years. This is equivalent to the corrosion rate of steel of about 0.3 mm/year.

Comparative case study

The rate of corrosion of the non-painted steel surface at the critical position is assumed to be 0.3 mm/year. For paint-coated steel, it may be natural to assume that at the beginning, the corrosion rate can be regarded as zero. However, at the final stage when the coating is about to disappear, the rate may be considered to be equal to that of the naked steel. Furthermore, we made an assumption that the annual acceleration rate of corrosion is $0.3/T$ mm/year^2, where T refers to the lifetime of the paint. Thus, for a given arbitrary time t, the annual incremental rate of corrosion is obtained by the integration as $0.3\,t/T$ mm/year. For example, at 5 years from the installation of the painted-steel rolled-H section, the annual incremental corrosion rate is $5 \times 0.3/T = 1.5/T$ mm/year and at the final year, namely at T years, the rate is $T \times 0.3/T = 0.3$ mm/year. Each of the rolled H-members is simply assumed to be subjected to the compressive load acting at the top end.

Basic comparisons are made among eight different cases with variations of interval for repainting:

1   Without any application of paint (Case 1), that is, naked steel
2   Polyurethane resin coating with 20-year interval of painting: Cases 2 and 3 where in the former, the repainting starts after 20 years and in the latter after 10 years
3   Fluorine resin coating: Cases 4, 5, 6, and 7 where the repainting starts after 30, 20, 10, and 10 years, respectively. In Cases 4 and 7, the interval of painting is 30 years; while in Cases 5 and 6, the interval is only 20 years
4   Protection with titanium lining: Case 8

In addition to the corrosion rate of steel taken as 0.3 mm/year, the life of a painted coating is assumed to be 20 years for the polyurethane resin, 30 years for the fluorine resin while the titanium lining is assumed to be maintained for more than 100 years. These eight cases are summarized in Table 8.9. The steel is assumed to cost JPY 700 thousand/ton and the painting cost is assumed to be JPY 7.2 thousand/m^2 initially and JPY5 thousand/m^2 for repainting in the case of polyurethane resin; JPY 7.4 thousand/m^2 initially and JPY 5.3 thousand/m^2 for repainting in the case of fluorine resin; while titanium costs JPY 100 thousand/m^2. For the least LCC at 100 years from the construction, the comparison study shows that the fluorine resin is the best with the repainting to be done at every 30 years and the member replaced once.

Figure 8.21 shows the deterioration of steel and paint coatings. The factor of safety against the yielding is prescribed to be 1.7. Thus, the allowable load on the rolled-H section is $17,500 \times 240 /1.7 N = 2,471$ kN. Assumption is

Table 8.9 Basic cases for comparison of coating methods

Case No.	(Initial cost; repainting cost) (in JPY/m^2)			
	No Paint	Polyurethane (7,200; 5,000)	Fluorine (7,400; 5,300)	Titanium-clad (100,000; 0)
1	x			
2		x		
3		x		
4			x	
5			x	
6			x	
7			x	
8				x

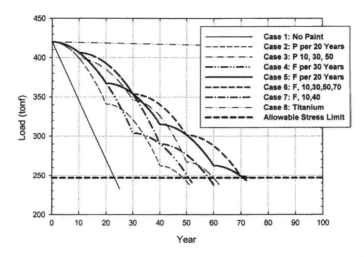

Figure 8.21 Comparisons of life among various corrosion protection cases.

made that the H-section must be replaced by a new member whenever the load-carrying capacity becomes less than the allowable stress. For example, in Case 4, the rolled-H member should be replaced once. Thus, it should be painted fresh twice and repainted two times too. Table 8.10 and Figure 8.22 show the corresponding costs for corrosion protection. When Figures 8.22 and 8.21 are compared, it is interesting to know that the strategy of extending the maximum life of the structural member does not necessarily lead to the most economical case.

Comparative study under variable unit price

Based on the results of these basic cases, further comparisons are made by considering possible reduction of painting cost by 10%, 20%, and 30% for the Polyurethane-resin coating, increase of painting cost by 10%, 20%, and 30% for the Fluorine-resin coating and reduction of cost for the titanium-clad steel by 20% and 50%. The cases are designated number 1–18 for various combinations of reduction and increment in the cost as listed in Table 8.11. Furthermore, Case I-J indicates the reduction or increase by $10 \times J\%$ on the basic Case I in Table 8.9.

By comparing the results shown in Table 8.12, the most reasonable case is found to be Case 4 where the fluorine-resin coating is applied at the painting interval of 30 years (cost = JPY2,194×1,000). This is followed by Case 3-3 when the unit price of the polyurethane-resin coating is reduced by 30% (cost = JPY2,236×1,000). However, it is amazing to note that there is not too much difference between Case 4-1 (cost = JPY2,243×1,000) and

Table 8.10 Comparison of cost among various paint coatings for rolled-H section

Case No.	Cost of steel 700/ton* (JPY1,000)	Cost of paint **polyurethane Initial: 7.2/m² Repaint: 5/m² (JPY1,000)	Cost of paint fluorine** Initial: 7.4/m² Repaint: 5.3/m² (JPY1,000)	Cost of titanium 100/m² (JPY1,000)	Total cost (JPY1,000)
1	$700 \times 1.215 \times 5$				4,253
2	$700 \times 1.215 \times 3$	$19.4 \times (7.2 \times 3 + 5 \times 4)$			3,354
3	$700 \times 1.215 \times 2$	$19.4 \times (7.2 \times 2 + 5 \times 5)$			2,465
4	$700 \times 1.215 \times 2$		$19.4 \times (7.4 \times 2 + 5.3 \times 2)$		2,194
5	$700 \times 1.215 \times 2$		$19.4 \times (7.4 \times 2 + 5.3 \times 4)$		2,400
6	$700 \times 1.215 \times 2$		$19.4 \times (7.4 \times 2 + 5.3 \times 5)$		2,502
7	$700 \times 1.215 \times 2$		$19.4 \times (7.4 \times 2 + 5.3 \times 4)$		2,400
8	700×1.215			19.4×100	2,791

Notes

* Total weight = 1.215 ton (12.15 kN); ** Total painting area = 19.4 m².

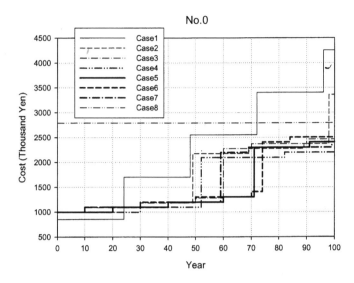

Figure 8.22 Comparison of LCC among various corrosion protection cases.

Table 8.11 Variation of unit price *f* from the basic cases: initial cost and repainting cost

Case No.	(Initial cost; repainting cost) (in JPY/m^2)		
	Polyurethane	Fluorine	Titanium-clad
	(7,200; 5,000)	(7,400; 5,300)	(100,000)
2-1	−10% (6,480; 4,500)		
3-1	−10% (6,480; 4,500)		
2-2	−20% (5,760; 4,000)		
3-2	−20% (5,760; 4,000)		
2-3	−30% (5,040; 3,500)		
3-3	−30% (5,040; 3,500)		
4-1		+10% (8,140; 5,830)	
5-1		+10% (8,140; 5,830)	
6-1		+10% (8,140; 5,830)	
7-1		+10% (8,140; 5,830)	
4-2		+20% (8,880; 6,360)	
5-2		+20% (8,880; 6,360)	
6-2		+20% (8,880; 6,360)	
7-2		+20% (8,880; 6,360)	
4-3		+30% (9,620; 6,890)	
5-3		+30% (9,620; 6,890)	
6-3		+30% (9,620; 6,890)	
7-3		+30% (9,620; 6,890)	
8-2			−20% (80,000)
8-5			−50% (50,000)

Table 8.12 Total cost in comparison: unit-price reduction or increment assumed

Case No.	Cost of steel 700/ton* (JPY1,000)	Cost of paint polyurethane** Initial: 5.04–6.48/m² Repaint: 3.5–4.5/m² (JPY1,000)	Cost of paint fluorine** Initial: 8.14–9.62/m² Repaint: 5.83–6.89/m² (JPY1,000)	Cost of titanium 50–80/m² (JPY1,000)	Total cost (JPY1,000)
2-1	700 × 1.215 × 3	19.4 × (6.48 × 3 + 4.5 × 4)			3,278
3-1	700 × 1.215 × 2	19.4 × (6.48 × 2 + 4.5 × 5)			2,389
2-2	700 × 1.215 × 3	19.4 × (5.76 × 3 + 4 × 4)			3,197
3-2	700 × 1.215 × 2	19.4 × (5.76 × 2 + 4 × 5)			2,312
2-3	700 × 1.215 × 3	19.4 × (5.04 × 3 + 3.5 × 4)			3,117
3-3	700 × 1.215 × 2	19.4 × (5.04 × 2 + 3.5 × 5)			2,236
4-1	700 × 1.215 × 2		19.4 × (8.14 × 2 + 5.83 × 2)		2,243
5-1	700 × 1.215 × 2		19.4 × (8.14 × 2 + 5.83 × 4)		2,469
6-1	700 × 1.215 × 2		19.4 × (8.14 × 2 + 5.83 × 5)		2,582
7-1	700 × 1.215 × 2		19.4 × (8.14 × 2 + 5.83 × 4)		2,469
4-2	700 × 1.215 × 2		19.4 × (8.88 × 2 + 6.36 × 2)		2,292
5-2	700 × 1.215 × 2		19.4 × (8.88 × 2 + 6.36 × 4)		2,539
6-2	700 × 1.215 × 2		19.4 × (8.88 × 2 + 6.36 × 5)		2,663
7-2	700 × 1.215 × 2		19.4 × (8.88 × 2 + 6.36 × 4)		2,539
4-3	700 × 1.215 × 2		19.4 × (9.62 × 2 + 6.89 × 2)		2,341
5-3	700 × 1.215 × 2		19.4 × (9.62 × 2 + 6.89 × 4)		2,609
6-3	700 × 1.215 × 2		19.4 × (9.62 × 2 + 6.89 × 5)		2,743
7-3	700 × 1.215 × 2		19.4 × (9.62 × 2 + 6.89 × 4)		2,609
8-2	700 × 1.215 × 2			19.4 × 80	2,402
8-5	700 × 1.215 × 2			19.4 × 50	1,820

Notes
* Total weight = 1.215 ton (12.15 kN); ** Total painting area = 19.4 m².

Case 4-2 (cost = JPY2,292×1,000) where the original unit price of the fluorine has been increased by 10% and 20%, respectively. Following next is Case 3-2 of polyurethane resin (cost = JPY2,312×1,000), Case 4-3 of fluorine resin (cost = JPY2,341×1,000) and Case 3-1 of polyurethane resin (cost = JPY2,389×1,000) where the unit price is reduced by 20%, increased by 30% and reduced by 10%, respectively. Finally, although it may be thought impractical to assume in Case 8-5 that the unit price of the titanium is reduced to one half (cost = JPY1,820 × 1,000), the advantage of not needing any replacement of the steel is promising.

Figures 8.23, 8.24, and 8.25 show for example LCCs for Case 8-5, Case 3-3, and Case 4-3; Case 8-5, Case 3-2, and Case 4-1; Case 8-2, Case 3-1, and Case 4-2, respectively assuming the reduction or increment of unit prices.

8.11   Concluding remarks

Most infrastructures have been constructed in such a way that they are economically feasible initially and in general supported by a sufficient budget. However, to present, things have changed rapidly. The infrastructure is deteriorating daily and the ordinary method of maintenance does not function well. The administrators, nevertheless, are finding increasing difficulty in securing the necessary budgets for repairs. Thus, it becomes necessary to

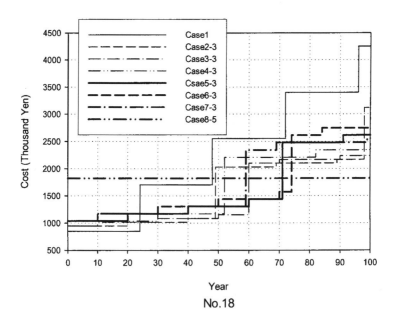

Figure 8.23 Life cycle costs corresponding to Case 8-5, Case 3-3, and Case 4-3.

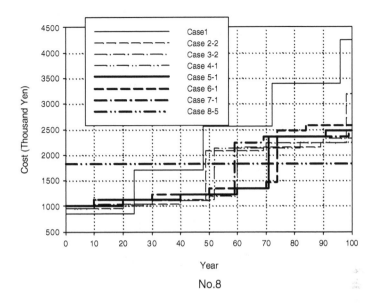

Figure 8.24 Life cycle costs corresponding to Case 8-5, Case 3-2, and Case 4-1.

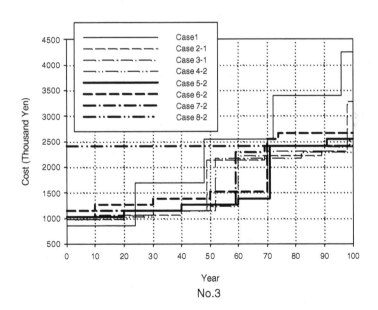

Figure 8.25 Life cycle costs corresponding to Case 8-2, Case 3-1, and Case 4-2.

provide reasonable maintenance plans and to implement them with the use of a LCC and LCM. In case of the need for an additional budget for maintenance, appropriate accountability is mandatory. For this purpose, the AMS will serve as a powerful tool in providing the necessary information for making decisions as to when and how the repair work should be carried out based on this fast-developing knowledge of maintenance. Thus, incessant efforts to update these systems are important.

Recently there has been positive progress in analyzing the electro-chemical corrosion-forming process and to improve the durability of structures since they are now designed to survive for 100 years or more; this improvement has been brought about by considering the LCC within the framework of the initial performance-based design and incorporating the results of exposure tests that are being made available and discussed more often, together with the appearance of new advanced structural materials.

In this chapter, the following conclusions may be drawn, first, there are various methods of corrosion protection available nowadays including organic lining, inorganic lining, petrolatum lining, paint coating, and cathodic protection. Thus, similar comparative studies such as those described previously using for example, heavy-duty salt-water resistant stainless steel in addition to titanium are strongly recommended to be performed in order to attain the best LCC.

Secondly, the corrosion-protection methods through paint coatings and titanium-clad steels for steel structures, that are built in the coastal areas of Japan, have been discussed in detail. It has been found that the most severe corrosion occurs at the splash zone or at the zone immediately below the MLWL. Based on a long 20-year corrosion-exposure test conducted on a rolled-H carbon-steel member, one obtains information on how to predict the LCC considering the corrosion rate due to deteriorating properties of paint coatings including polyurethane resin and fluorine resin. From the simulated results, a better LCC may be obtained by using the Fluorine-resin coating than by using the Polyurethane-resin coating because of the longer durability of the former resin, namely, 30 years of painting life as compared with only 20 years of the latter type of resin.

Thirdly, the use of titanium-clad steel may also become competitive, for example, if the commercial price for titanium-clad steel could be reduced by 50% from the present price. Moreover, not only titanium, but the durability of heavy-duty saltwater-resistant stainless lining has been studied in the past decade by fabricators and may also be applied advantageously for the protection of structural members in the splash zone.

Acknowledgments

The authors wish to express their thanks to members of the Bridge Asset Management Teams of Aomori Prefecture, Hanshin Expressway Company

Limited and the Study Group on Durability and Corrosion protection of Steel Structures of the Japan Iron and Steel Federation (JISF) for useful discussions. This study was financially supported by the JISF.

References

Abe, M. "On Elongation of Life of Steel Tubular Piles under Offshore Environment-Long-term Corrosion Exposure Tests at Hazaki Marine Observatory Wharf," *Gekkan Kensetsu* (Construction Monthly), pp. 26–28, February 2001 (in Japanese).

Corotis, R.B., Schueller, G.I. and Shinozuka, M. (ed.), ICOSSAR '01, *Structural Safety and Reliability*, A.A. Balkema Publishers, 2001.

Division of Road Asset Management System, *Annual Report*, Hanshin Expressway Public Corporation, 2003 (in Japanese).

Dunker, K.F. and Rabbat, B.G., "Why America's Bridges Are Crumbling," *Scientific American*, pp. 66–72, March 1993.

Floating Bridge Committee, Guidelines for Design of Floating Bridges, Steel Structures Series 13, JSCE, March 2006.

Frangopol, D.M. and Parag C. Das, "Management of bridge stocks based on future reliability and maintenance costs," *Current and future trends in bridge design, construction and maintenance, Thomas Telford*, Singapore, pp. 45–58, 1999.

Frangopol, D.M., Jung S. Kong, and Emhaidy S. Gharaibeh, "Reliability-based Life-cycle Management of Highway Bridges," *Journal of Computing in Civil Engineering*, Vol. 15, No. 1, January, ASCE, pp. 27–34, January 2001.

Japan Association of Steel Bridge Construction, JASBC, "Guide Book for Paint Coating for Bridge Engineers," March 2000 (in Japanese).

Japan Association of Steel Bridge Construction, JASBC, "Revised Manual for Management of Attached Salt for Steel Bridges," March 2001 (in Japanese).

Japan Association of Steel Bridge Construction, JASBC, "Life Cycle Cost of Steel Bridges (Revised) & Its Manual," October 2001 (in Japanese).

Japan Association of Steel Bridge Construction, JASBC, "Q & A on Anti-corrosion of Steel Bridges," March 2002 (in Japanese).

Japan Road Association, "Specifications and Commentaries on Roadway Bridges," March 2002 (in Japanese).

Japanese Association of Port and Harbor, JAPH, "Technological Standard and Commentary of Port and Harbor Facilities," Vols 1 & 2, 1999 (in Japanese).

JSSC, "For Reduction of LCC of Paint Coating on Steel Bridges," August 2002 (in Japanese).

JSSC, "LCC Assessment and Anti-Corrosion Design of Steel Bridges," Anti-Corrosion and LCC Subcommittee, Committee for Study of Performance-based Design for Steel Bridges, JSSC, September 2002 (in Japanese).

JTS, "Titanium World – In pursuit of its unlimited potentialities and realization of dreams," The Japan Titanium Society, http://www.titan-japan.com, No. 7, October 2000 (in Japanese).

Kaneuji, M., Asari, H., Takahashi, Y., Ohtani, Y., Ukon, H. and Kobayashi, K., "Development of BMS for a Large Number of Bridges," *Bridge Maintenance,*

Safety, Management, Life-Cycle Performance and Cost, IABMAS, Taylor and Francis, pp. 597–598 (abstract) and CD (full text), 2006.

King, M.J., Howells, J.D., and Sanders, P.A., "Songsu Bridge, South Korea, Widening of an Existing Multi-span Truss bridge, Concept & Design," *Current and Future Trends in Bridge Design, Construction and Maintenance, Thomas Telford,* pp. 257–270, Singapore, December 1999.

Kozai Club (The Japan Iron and Steel Club), "*Q & A on Anti-corrosion for Offshore Structures,*" Gihou-do, 2001 (in Japanese).

Masahiro Yamamoto, K. Yoshida, and N. Hirosawa, "Analysis of Corrosion Process of a Steel Member Exposed in the Ocean," *Proceedings of Annual Meeting of JSCE,* V-160, Vol. 46, September 1991, pp. 336–337 (in Japanese).

Masahiro Yamamoto, K. Yoshida, and T. Imai, "Analysis of Corrosion Process of a Steel Member used for a Long Period in the Ocean," *Proceedings of Annual Meeting of JSCE,* V-184, Vol. 47, September 1992, pp. 398–399 (in Japanese).

Masahiro Yamamoto, N. Hirosawa, K. Yoshida, C. Kato, and T. Hada, "Characterization of Surface Texture of Corroded Steel Plates Exposed in Marine Environment," *Zairyo to Kankyo (Materials and Environment),* Vol. 41, 1992, pp. 803–808 (in Japanese).

Masubuchi, S. and H. Yokota, "Life-Cycle Cost Analysis of Berthing Facilities and Development of a Decision Support System during their Maintenance Work," *Report of the Port and Harbour Research Institute,* PHRI, Vol. 38, No. 2, June 1999 (in Japanese).

Matsumura, E., Senoh, Y., Sato, M., Miyahara, Y., Kaneuji, M., and Sakano, M., "Condition Evaluation Standards and Deterioration Prediction for BMS," *Bridge Maintenance, Safety, Management, Life-Cycle Performance and Cost, IABMAS,* Taylor and Francis, pp. 597–598 (abstract) and CD(full text), 2006.

Muranaka, A.O. Minata, and K. Fujii, "Estimation of Residual Strength and Surface Irregularity of the Corroded Steel Plates," *Journal of Structural Engineering,* JSCE, Vol. 44A, pp. 1063–1071, March 1998 (in Japanese).

Nippon Steel, "Titanium Products for Building Construction," Cat. No. TC021, June 2000 (in Japanese).

Nippon Steel, "Titanium Clad," Cat. No. TC026, November 2000 (in Japanese).

Nishibayashi, M., Kanjo, N., and Katayama, D., "Toward More Practical BMS: Its Application on Actual Budget and Maintenance Planning of a Large Urban Expressway Network in Japan," *Bridge Maintenance, Safety, Management, Life-Cycle Performance and Cost, IABMAS,* Taylor and Francis, Porto, Portugal, pp. 915–916 (abstract) and CD(full text), 2006.

P. Thoft-Christensen, "Estimation of Bridge Reliability Distributions," *Current and Future Trends in Bridge Design, Construction and Maintenance, Thomas Telford,* pp. 15–25, Singapore, December 1999.

Port and Airport Research Institute (PARI), Coastal Development Institute of Technology (CDIT) and Japanese Association of Steel Pipe Piles (JASPP), Study on Corrosion Protection Methods Applicable to Offshore Steel Structures— Long Period Exposure Tests—June 2006 (a brochure) (http://www.cdit.or.jp and http://www.jaspp.com).

Public Work Research Institute (PWRI) and Japan Iron and Steel Federation (JISF), Study on Corrosion Protection for Steel Structures at Splash, Ebb and Flow and Underwater Zones, July 2006 (a brochure).

Sakai, K., "Goal of Performance-based Design," Research and Development of Bridges, Special Issue on New Bridge Technology for 21st Century, Bridge and Foundation, Vol. 31, No. 8, pp. 73–83, 1997 (in Japanese).

Sato, H., et al., "Development of Metallic Sheathing for Protection of Offshore Steel Structures Using Seawater Resistance Stainless Steel," Nippon Steel Technical Report, No. 87, January 2003.

Takagi, S., "Present State of Bridge Maintenance and Management of USA and its Challenges," Special Issue on Maintenance and Reclamation Infrastructure, *Journal of JSCE*, Vol. 85, pp. 52–54, 2000 (in Japanese).

Urban Expressway Study Group, *Manual for Inspection and Repair for Road Bridges in Urban Expressways*, Hanshin Expressway Public Corporation, Rikoh Tosho, October 2004 (in Japanese).

Watanabe, E., "Floating Bridges: Past and Present," *Structural Engineering International*, Vol. 13, No. 2, pp. 128–132, 2003.

Watanabe, E., T. Maruyama, Y. Kawamura, and H. Tanaka, "Why is a Floating Swing Arch Bridge Built in the Port of Osaka?" *Proceedings of New York City Bridge Conference, Journal of Bridge Engineering*, ASCE, November 2001.

Watanabe, E., Sugiura, Utsunomiya, T., and Yamamoto, M. "Residual Structural Performance of Rolled H Members Submerged in Seawater for a Long Time and Their Anti-corrosion Strategy," *Bridge Maintenance, Safety, Management, Life-Cycle Performance and Cost, IABMAS*, Taylor and Francis, Porto, Portugal, pp. 207–209 (abstract) and CD (full text), 2006.

Yamamoto, T.M. Oda, T. Morita, Y. Ishihara, and K. Higo, "Prediction of Service Life on Heavy Duty Coating for Marine Steel Structures," *Boshoku Gijutsu (Anti-corrosion Technology)*, Vol. 35, pp. 3–9, 1986 (in Japanese).

Yamamoto, N., Kaneuji, M. and Watanabe, E. "Implementation of Bridge Management System in Aomori Prefectural Government," *Bridge Maintenance, Safety, Management, Life-Cycle Performance and Cost, IABMAS*, Taylor and Francis, Porto, Portugal, pp. 599–600 (abstract) and CD (full text), 2006.

Chapter 9

Research and development of VLFS

Hideyuki Suzuki

9.1 Introduction

Pontoon-type very large floating structures (VLFS) comprises pontoon hulls joined together and designed for operation in protected waters. It is further characterized by hydroelastic behavior as described in the preceding chapters. Moreover, due to the unprecedented length scales and the geometrical configurations involved, designers must expect to contend with unfamiliar potential global failure modes. These VLFS characteristics pose engineering challenges in designing VLFS.

This chapter deals with the research and development of VLFS, especially on the Mega-Float in Japan. The applications and history of pontoon type VLFS are also presented.

9.2 Applications and history of pontoon-type VLFS

Interest in utilizing the space afforded by the seas surrounding a nation, for purposes other than conventional shipping or ocean-resource extraction, has increased as coastal-population density increases. Until the potential of modern shipbuilding technology became apparent in the 1950s, the only manner in which this ocean space could be exploited on a large scale was through land reclamation. Such exploitation is, however, limited to shallow regions of the continental shelf.

The first concept of VLFS that appeared in the modern world after the industrial revolution was the Floating Island described by the nineteenth-century French novelist Jules Verne, one of the founders of science fiction. The first VLFS promoted in earnest was the Armstrong Seadrome. It was proposed initially to enable airlines to cross the world's oceans (Armstrong 1924). Its stability was demonstrated in tank tests, and various other related platforms were promoted until 1955 (Nelson 2001).

In the 1950s, architects were drawn to the idea of floating cities and such a concept was demonstrated in part at the Okinawa International Ocean

Table 9.1 Milestone developments in VLFS technology

1924–1955	Armstrong Seadrome and related concepts
1950s	Floating-city concepts in architecture and urban design
1960s	Puppet drama "Hykkori Hyoutan Jima"
1973–1974	Proposal for floating airport for Kansai International Airport Phase 1 construction, semi-submersible-type floating structure
1975	Okinawa International Ocean Exhibition – Aquapolis
1988	Kamigoto Oil Stockpile 390 m × 97 m × 27.6 m × 5 Units
1994	Proposal for floating runway for Kansai International Airport Phase 2 construction, pontoon-type floating structure
1995	Technological Research Association of Mega-Float (TRAM 1999a)
1995–1996	TRAM Phase1 Experiment 300 m × 60 m × 2 m (TRAM 2001)
1996	Shirashima Oil Stockpile 397 m × 82 m × 25.1 m × 8 Units
1997–2001	TRAM Phase 2 Experiment 1,000 m × 60–120 m × 3 m Landing & takeoff experiments (TRAM 2002)
2001–2005	Research and development by the Shipbuilding Research Center. Proposed Haneda International Airport Runway; pontoon/semi-submersible combination hull

Source: Proceedings of the 16th International Ship and Offshore Structures Congress, Vol. 2.

Exhibition in 1975 with the construction of a semi-submersible unit of such a city. In a similar manner, a floating airport was proposed for the new Kansai International Airport in 1973. Since the early 1970s, the technology for very large floating structures has gradually developed, while changing societal needs have resulted in many different applications of the technology being considered.

Milestones in the development of VLFS are listed in Table 9.1. A big milestone other than application-related developments is the founding of the Technological Research Association of Mega-Float (TRAM) in 1995. A mega-float is defined as a pontoon-type VLFS, which includes both mooring and access systems. It is intended for deployment in protected waters. Fundamental design, construction and environmental impact-related technologies were developed and overall safety assessment was carried out with a budget of USD172 million (TRAM 1999a–2002).

9.2.1 Airports

Proposals to use floating structures for airports were first considered in the 1920s to enable airplanes to cross the world's oceans. These concepts were investigated more seriously for military applications by the United States in the 1940s. The US Navy Civil Engineering Corps developed a floating pontoon flight deck in the early 1940s for use by Great Britain. It comprises many hinged arrays of pontoons, and measured 552 m × 83 m × 1.5 m with a 0.5 m draft, and was deployed in protected water. Takeoffs and landings of aircrafts were successfully demonstrated in 1943 (Laycock 1943).

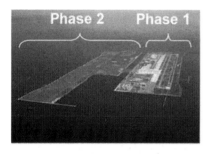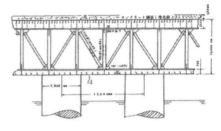

Figure 9.1 Kansai International Airport and floating structure proposed for Phase 1 construction.

Source: Shipbuilding Research Center of Japan.

With advanced improvements in VLFS technology, a floating airport was proposed for the new Kansai International Airport in Japan in 1973. A floating airport was initially proposed for Phase 1 construction. Though the proposal was ultimately declined in favor of land reclamation, the industry had formally commenced research on VLFS technology. Figure 9.1 shows the Kansai International Airport and the concept of a floating structure proposed in 1973. Although the initial phase was not built as a floating structure, interest in the concept remains strong. This is the area of VLFS research that has received the most attention, due in a large part to the efforts of the TRAM which was active in Japan from 1995 to 2001.

9.2.2 Offshore port facilities

Floating offshore port facilities are an attractive proposition in regions where suitable land close to urban centres is rather limited. Proposals have been produced for offshore container terminals to service large ocean-going vessels and supply the immediate hinterland with feeder container ships. It may also be beneficial to site terminals for potentially hazardous vessels, such as LNG carriers, offshore. Figure 9.2 shows the concept of a floating container terminal.

9.2.3 Offshore storage and waste disposal facilities

The potential of a VLFS as a storage facility is demonstrated in Japan. The Japanese government had decided to construct ten national oil stockpiles, based on the lesson learned from the oil crises experienced in 1973 and 1979. Two of the ten bases were selected to be on a floating base. The Kamigoto oil stockpile was constructed at the Kamigoto island of Nagasaki

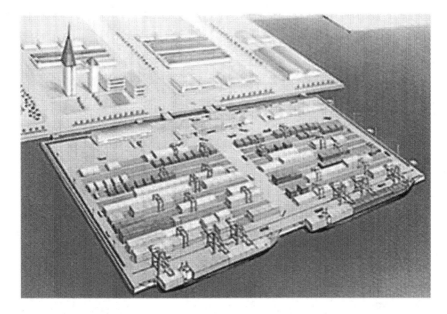

Figure 9.2 Offshore container terminal.
Source: Shipbuilding Research Center of Japan.

Figure 9.3 Kamigoto oil stockpile.
Source: Shipbuilding Research Center of Japan.

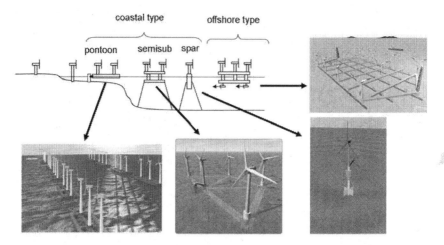

Figure 9.4 Various concepts of floating wind turbines.
Source: Yato et al. (2003), Suzuki et al. (2004), and Inoue et al. (2004).

in 1988. Figure 9.3 shows the Kamigoto oil stockpile. The oil stockpile consists of five floating oil storage barges that measure 390-m long, 97-m wide, and 27.6-m deep. The Shirashima oil stockpile that has seven oil storage barges, each measuring 397-m long, 82-m wide, and 25.1-m deep, was constructed at the Shirashima Island, off the coast of Fukuoka City in 1996.

In increasingly densely populated coastal regions, the ability to site storage facilities of any kind, together with waste-processing and treatment plants, out of sight of land is an attractive one and such a facility may also incorporate power-generation capability.

9.2.4   Energy islands and food production

In an extension to merely generating power from waste disposal, an offshore facility may be considered for housing a range of sustainable-energy technologies. Depending on the prevailing climate, such a structure may include some, or all, of wind turbines, wave-power generators, tidal-current turbines, and ocean-thermal energy-conversion units. Such a structure may also be a natural host to environmental research activities and food production through aquaculture and marine biomass plantations. Variations on this application of very large floating structure technology are being proposed by researchers and engineers from Japan, France, the United Kingdom, the United States, and South Korea. Figure 9.4 shows the various concepts of floating wind turbines (Yago et al. 2003; Inoue et al. 2004; Suzuki 2004).

Figure 9.5 Aquapolis in Okinawa International ocean exhibition.

Source: Shipbuilding Research Center of Japan.

9.2.5 Habitats

Shipbuilding technology had attracted the attention of Japanese architects in the late 1950s, and there was a movement in architecture and urban design to utilize ocean space and expand human habitation onto the ocean surface (Kikutake 1994).

The floating-city project started in the University of Hawaii in 1971. Aquapolis, a large semi-submersible shown in Figure 9.5, portrayed a unit of a floating-city concept, and was constructed for the Okinawa International Ocean Exhibition held in 1975.

As the original idea for a VLFS was habitat, it is perhaps surprising that more plans for offshore floating cities have not been developed over the years, although there are current proposals for offshore sports facilities and theme parks in Japan and South Korea. However, with the ever-increasing pressure on coastal zones from urban populations and the threat of environmental change, it is likely that such ideas will again become the focus of attention in the coming years, taking advantage of the technology already developed and providing impetus for future research.

9.3 Technological Research Association of Mega-Float

The Transport Technological Council of the Ministry of Land, Infrastructure, and Transportation recommended the promotion of a VLFS in 1993.

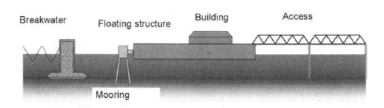

Figure 9.6 Concept of Mega-Float.

Based on this recommendation, the Technological Research Association of Mega-Float was founded in 1995. The association carried out research and development on a VLFS for six years, which terminated in 2001 (TRAM 1999a, 2000, 2001). The concept pursued by this association was Mega-Float, a pontoon-type floating structure, which is more cost-effective, competitive and suitable for development in protected waters such as in a large bay. The concept of Mega-Float is shown in Figure 9.6. Mega-Float consists of a floating structure, a mooring system and access infrastructure. If necessary, breakwater construction was considered. The general research objectives of the association were to prove and demonstrate the soundness of the technology.

The research and development programme was divided into two phases. The fundamental technologies of design and construction were investigated and established. Design guidelines were compiled and legal aspects of Mega-Float construction in Japan were studied. Onsite experiments that used both a 300-m-long Phase 1 model and a 1,000-m-long Phase 2 model were conducted to demonstrate the soundness of the technology. Takeoff and landing experiments were conducted during Phase 2.

Possible applications of the Mega-Float are broad. Examples include floating airports, offshore container terminals, floating sports facilities, leisure facilities, waste processing facilities, and floating emergency-rescue bases. The greatest importance was placed on research on the floating airport and runway. In the Tokyo area, the annual numbers of passengers are 56.4 million for the Haneda International Airport and 27.4 million for the Narita International Airport, while air-freight capacity is insufficient to serve the economy of the area. The Tokyo Metropolitan Third Airport and a new runway for Haneda International Airport were long-awaited expansion plans. Budgets and activities of the association are shown in Table 9.2 (Sato 2003).

The association terminated its formal activities in 2001, which were succeeded by the Shipbuilding Research Center of Japan (SRCJ) and the Shipbuilders' Association of Japan, and a new runway for the Haneda International Airport was designed. Some projects with small-scale floating structures and reused structures of Phase 2 models were planned and put into practice.

Table 9.2 Budgets and schedules of the association

	Phase I (1995–1997)	Phase 2 (1998–2001)
Objective	Establish basic technology	Establish airport construction technology
Experiment	300-m-long model; joining of units at sea	1,000m-long model; joining of units at sea
Research	Design fabrication and joining at sea	ILS test
		Landing and takeoff of airplane
	Operational requirement	Concept study
	Environmental impact	Legal aspect
Budget	USD68.2 million	USD103.6 million

Source: Sato (2003).

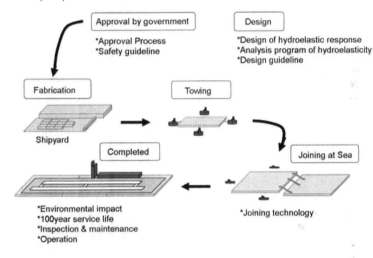

Figure 9.7 Realization process of Mega-Float.

9.3.1 Research of TRAM

The research of TRAM is broad but may be categorized into five areas that follow the flow of the realization process of Mega-Float. These research areas are design, government approval, construction, safety and serviceability, and environmental impact.

Figure 9.7 shows the process of realization of Mega-Float and the research areas focused upon by TRAM.

Analysis and design tools for hydroelastic behavior

The dynamic behavior of the Mega-Float is characterized by its hydro-elastic behavior. Several hydroelastic computer programs were developed with various complexities and levels of modeling, both inside and outside

Table 9.3 Analysis programs developed for global response

Program	A	B	C	D	E
Fluid domain	2-D domain decomposition	3-D domain decomposition	3-D domain decomposition	3-D boundary element method and finite element method	3-D domain decomposition and finite element method
Water depth	uniform	uniform	uniform	uniform	variable
Draft	uniform	uniform	uniform	uniform	variable
Structure	beam	plate	plate	finite-element method	finite-element method
Shape		rectangular	combination of rectangular	arbitrary	arbitrary
Stiffness	uniform	uniform	uniform	variable	variable
Mass	uniform	uniform	uniform	variable	variable
Break-water			considered		considered

Source: TRAM (1999a, 2000, 2001).

TRAM (TRAM 1999a, 2000, 2001). Some of the programs are capable of only evaluating a global hydroelastic response and are used for serviceability and general safety requirement studies. Others are developed for use in detailed structural design. In Table 9.3, some of the analysis programs developed for global-response analysis are presented. Program E in the table is developed for structural analysis capable of analyzing detailed structural analysis. Figure 9.8 shows the analysis procedure of the so-called "two-step" method that was developed for the structural analysis of Mega-Float and the programs.

Structural safety

The structural behavior of Mega-Float due to a plane's belly landing, the impact of a falling jet engine nacelle and crash of an airplane onto the Mega-Float was investigated. The most severe case simulated was the vertical fall and crash of the fuselage of a B747 jet onto the floating structure. The assumed weight of the fuselage was 500 t and its speed was 100 m/s. It was confirmed that the damage to the Mega-Float was limited. Figure 9.9 shows the simulation results from a vertical fall of the fuselage.

The effects of an earthquake were also investigated. Although the horizontal motion of the sea bottom due to an earthquake is transmitted through the

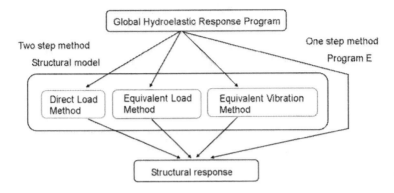

Figure 9.8 Analysis program and procedure for the structural analysis.

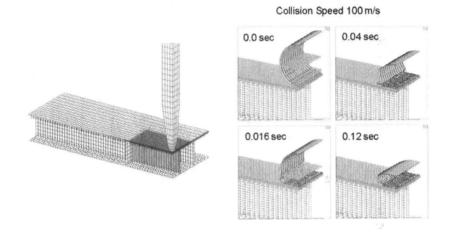

Figure 9.9 Simulation model and results crash of fuselage.

mooring dolphins and rubber fenders to the floating structure, the floating structure may be practically regarded as isolated from the effects of an earthquake. The impact of the earthquake was generally limited to the mooring system. When the seismic center was close to the installation site and the vibration was transmitted vertically through a water column, the impact to the structure may not be as negligible.

Design criteria

Serviceability and safety criteria were key issues in the design of the Mega-Float. Design-criteria development for the Mega-Float was needed if it was to be used as an airport. For serviceability, existing standards and rules, such

Table 9.4 Typical serviceability criteria of Mega-Float

Facility	Criteria	Rule
Runway	Slope longitudinal < 1.0°, transverse < 1.5°, radius of curvature > 30,000 m	Airport facility design standard
Taxiway	Slope longitudinal < 1.5°, transverse < 1.5°, radius of curvature > 3,000 m	Airport facility design standard
ILS/GS	Misalignment < 0.144°	Civil aeronautics law
PAPI	Misalignment < 0.1°	Civil aeronautics law

Source: TRAM (1999a, 2000, 2001).

as the Airport Facility Design Standard, Civil Aeronautics Law, Building Standard Law, and the Standard for Cranes, were investigated. The influence of elastic responses on an Instrument Landing System (ILS), Precision Approach Pass Indicator (PAPI), and Future Air Navigation System (FANS) was investigated using an airline flight simulator. Fluctuations in angles of Glide Slope (GS) signals due to hydroelastic responses of runways were simulated and the influence on controllability of airplanes was investigated (TRAM 1999a, 2000, 2001). Table 9.4 shows the typical serviceability criteria of Mega-Float that were derived from the research.

Another research area was the technology that would guarantee a long, service life. If the Mega-Float was to be used as an airport, the service life would need to be 100 years. Maintenance systems of existing floating structures and experiences of national oil stockpiles were investigated. Applications of anti-corrosive materials, such as titanium cladding, painting and cathodic protection system, and monitoring and maintenance systems, were investigated (TRAM 1999a, 2000, 2001).

The Technical Guidelines of Mega-Float that summarizes the research findings was compiled. Table 9.5 shows the contents of the guidelines. Based on the guidelines, a 4,000 m generic airport, shown in Figure 9.10, was designed and this model was used for subsequent research (Sato 2003; TRAM 2000). One notable item in the guideline was the recommendation of risk-based evaluation for overall safety.

Overall safety of Mega-Float

With regard to the overall safety evaluation, scenarios of catastrophic failure of Mega-Float and acceptable risks were investigated (Fujikubo et al. 2003; TRAM 1998; Kato et al. 2001). Worst system-failure scenarios were screened from possible scenarios by expert investigation as shown in Figure 9.11. For acceptable safety criteria, safety levels of various activities in Japan were investigated and the fatal accidental rates of the activities were calculated. Qualitative and psychological factors that affected the

Table 9.5 Contents of technical guidelines for Mega-Float

Volume 1	General rules
Volume 2	Environmental impact assessment
Volume 3	Materials
Volume 4	Design load
Volume 5	Hull structures
Volume 6	Station keeping facility
Volume 7	Wave control facility
Volume 8	Disaster prevention measures
Volume 9	Quality control for construction workers
Volume 10	Maintenance and inspection
Volume 11	Overall safety evaluation

Source: TRAM (1999b).

acceptable safety criteria and the formula for the acceptable safety criteria proposed in the past were investigated. Trial calculations of target safety levels of Mega-Float, when used as an international airport, were made. Figure 9.12 shows the estimated target safety levels of a Mega-Float airport (Suzuki 2002).

Mega-Float was to be safeguarded against catastrophic failure, such as sinking, drifting, and the catastrophic collapse of the floating structure. The typical Mega-Float used for the illustrative safety assessment had an assumed deck area of 500 ha and was moored by more than 30 dolphins. The collapse behavior of floating structures, due to extreme wave conditions, was investigated. For progressive failures of the mooring system, the collapse behavior of a single dolphin was investigated by pushover analysis. Based on these results, the system failure was investigated. Mega-Float was to be installed at such a site where it is protected from possible tsunamis, typically a site inside a large bay or at some distance from the shoreline. The magnitude of a tsunami was estimated by numerical simulation and analysis conditions were determined. The collapse behavior and redundancy of the mooring system under these conditions was also investigated.

Collision analysis was performed that assumed a maximum ship size and speed that is statistically expected in a large bay, such as Tokyo Bay. Mega-Float is divided into a large number of watertight compartments. The effects of changes in the size of the compartment and the magnitude of damage on the remaining strength after the damage and flooding were investigated. From this research, the requirements for compartment division were determined.

Legal jurisdiction and government approval

The plan of Mega-Float was to be evaluated and approved by the authority that is responsible in overseeing the entirety or part of the Mega-Float. The government legal and approval process was investigated. The general plan of Mega-Float was to be compliant with both the Port and Harbor Law and

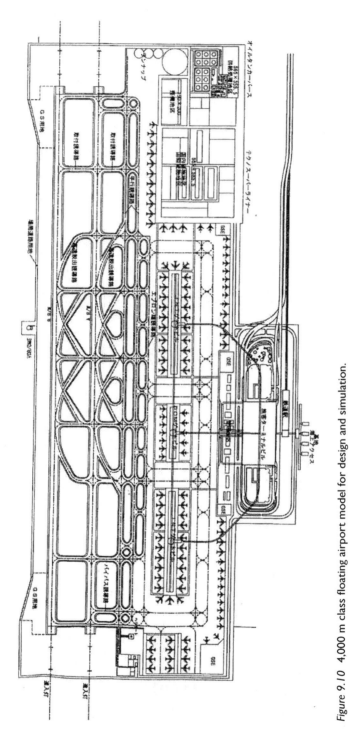

Figure 9.10 4,000 m class floating airport model for design and simulation.

Source: Shipbuilding Research Center of Japan.

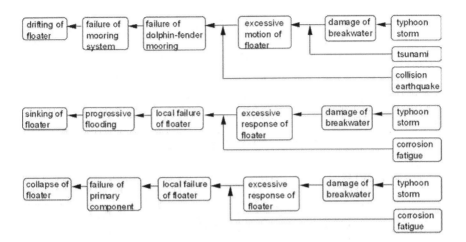

Figure 9.11 System failure scenario of Mega-Float.

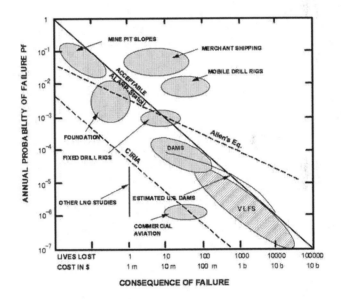

Figure 9.12 Target safety level tentatively evaluated for Mega-Float airport.

the Fishing Port Law. Buildings on the Mega-Float were to be regulated by both the Building Standard Law and the Fire Defenses Law. Floating Structures are regulated by the Ship Safety Law. Approval processes differ from law to law. A Mega-Float Safety Evaluation Committee was proposed and accepted by the government. Experts and all government bureaus in charge

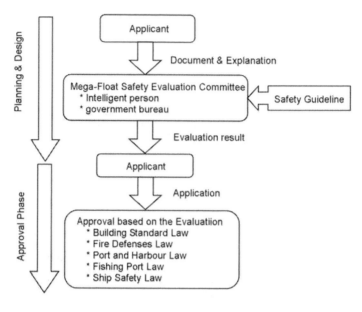

Figure 9.13 Approval process in government.

of the approval gathered in the committee and evaluated the application. Once the plan was judged to be acceptable, each bureau approves the plan. Figure 9.13 shows the process of government approval. The Safety Evaluation Guideline of Mega-Float, which forms the backbone of the committee, was investigated, proposed and accepted by the government bureaus (Committee on Realization of Megafloat 2001). The contents of the guideline are shown in Table 9.6. Although each elemental structure is designed by a well-established design rule, as mentioned above it is necessary to confirm that the whole system satisfies expected safety levels. A risk-based safety evaluation of the total system was included in the guideline.

Fabrication and towing of units

Units of the Mega-Float are simple box-like structures and construction itself is not a difficult task. Most of the technological developments in the construction phase were related to construction operations at sea. Experiments to test the towing of Mega-Float units were carried out during the construction of onsite experimental models (Hara et al. 2003). The effects of unit size and wave conditions on the strain induced in the towed unit were investigated. In Phase 2 experiments, a unit that measured 300-m long, 60-m wide, and

Table 9.6 Contents of the safety guidelines of Mega-Float

Volumes	Content
Volume 1	General rules
Chapter 1-1	General
Chapter 1-2	Fundamental concept for safety of VLFS
Volume 2	Materials
Chapter 2-1	General
Volume 3	Design load
Chapter 3-1	General rule
Chapter 3-2	Dead load
Chapter 3-3	Live load
Chapter 3-4	Environmental load
Chapter 3-5	Accidental load
Volume 4	Hull structures
Chapter 4-1	General rules
Chapter 4-2	Water-tightness and compartments
Chapter 4-3	Structural strength
Chapter 4-4	Preventive measures against material deterioration
Volume 5	Station keeping facility
Chapter 5-1	General
Chapter 5-2	Configuration, arrangement, and structural strength of station keeping facility
Volume 6	Superstructure
Chapter 6-1	General
Chapter 6-2	Arrangement and structure
Volume 7	Access facility
Chapter 7-1	General
Chapter 7-2	Structure
Volume 8	Disaster prevention measures
Chapter 8-1	General
Chapter 8-2	Disaster prevention control
Chapter 8-3	Disaster prevention planning
Volume 9	Quality control for construction works
Chapter 9-1	General
Chapter 9-2	Quality control
Volume 10	Maintenance and inspection
Chapter 10-1	General
Chapter 10-2	Maintenance and inspection
Volume 11	Overall safety evaluation
Chapter 11-1	General
Chapter 11-2	Evaluation of safety

Source: TRAM (1999c).

2-m deep was towed 195 miles in the Pacific Ocean from the Tsu works of Nippon Kokan to the experimental site of Yokosuka in Tokyo Bay. The numerical method for predicting the response of units in wave conditions was verified and a critical wave condition was identified during the operation. Towing resistance and drift force were also investigated (TRAM 2000).

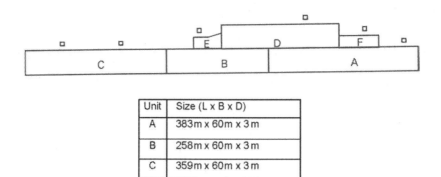

Unit	Size (L x B x D)
A	383 m x 60 m x 3 m
B	258 m x 60 m x 3 m
C	359 m x 60 m x 3 m
D	300 m x 60 m x 3 m
E	64 m x 31.3–34.5 m x 3 m
F	100 m x 29.70 m x 3m

Figure 9.14 Unit division of Phase 2 model.

Source: Shipbuilding Research Center of Japan.

Joining technology of units at sea

Mega-Float is constructed by joining unit structures that were fabricated in shipyards. Unit structures are fabricated in the well-controlled environments of shipyards but the joining of the units take place at sea and are exposed to the natural environment of the installation site. Construction with either dry welding with a water-draining device or wet welding at sea was investigated (Yamashita 2003; TRAM 2000). The influence of both wave conditions and unit-joining sequences on the responses of structure and performance of construction were investigated (TRAM 2000). The developed joining technology was tested in the construction of Phase 1 and Phase 2 experimental models. Figure 9.14 shows the unit division of the 1,000-m long Phase 2 model. Figure 9.15 shows the thermal deformation of the Phase 2 model due to sunlight. The estimated maximum deformation of the model was 190 mm.

Environmental impact study

A concern of the public in general for large-scale development projects like Mega-Float is the environmental impact. The influence on the ecosystem due to a possible block of flow and a large shaded space created below Mega-Float was investigated. Research on the environmental impact of Mega-Float was carried out to answer the questions presented by both experts and the public (homepage of SRCJ; TRAM 1999a–2001). The study involved the physical environment, such as flow, salinity, and temperature around the

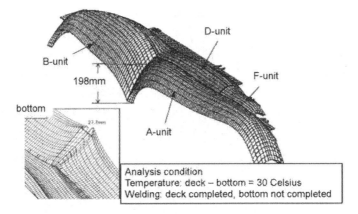

Figure 9.15 Deformation of Phase 2 model due to sunlight.

Source: Shipbuilding Research Center of Japan.

Mega-Float. Water circulation was investigated by numerical simulation and field measurements during the Phase 2 experiments. The ecosystem around the Mega-Float and the water quality around and below the Mega-Float were investigated by numerical models of the ecosystem and field measurements. Few changes were observed in various indices of either the physical or ecological environments, except for a small reduction in dissolved oxygen (DO) adjacent to the floating structure as a result of the activity of attached organisms.

9.3.2 Onsite experimental models

The Phase 1 model was constructed to investigate the basic technology of design, simulation, fabrication, joining at sea and environmental impact (Sato 2003). The size of the Phase 2 model, which is substantially the same size as the existing, small commuter airport, was chosen as the model size. Photos of the Phase 1 and Phase 2 models are shown in Figure 9.16. The design and construction of the Phase 2 model started in 1998 and onsite experiments began in 1999 (Sato 2003). The experiments were completed by the end of 2000 and the model was removed from the experimental site in 2001 (Sato 2003). Principal dimensions of the models are shown in Table 9.7. The research objectives of the Phase 2 model are

1 Verification of applicability of the simulation program of hydroelastic response, the simulation program of mooring, and the evaluation of construction technology
2 Research on airport facilities

Phase 1 Experimental model   Phase 2 Experimental model

Figure 9.16 Phase I and 2 models.

Source: Shipbuilding Research Center of Japan.

Table 9.7 Principal dimension of Phase I and 2 models

	Phase I model	Phase 2 model
Length	300 m	1,000 m
Breadth	60 m	60 m (partially 121 m)
Depth	2 m	3 m
Draft	0.5 m	1 m
Total area	1.8 ha	8.4 ha
Runway	—	900 m × 25 m

Source: Sato (2003).

3 Experiment on flight-navigation facilities by the Japanese Civil Aviation Bureau using checkers regarding ILS, GS, and FANS
4 Confirmation of safe landing on the model runway
5 Investigation of environmental impact organisms that lived under the model and seabed, in addition to the quality of sea water, were measured and evaluated. Environment-assessment procedures of Mega-Float were proposed.

The Phase 2 model was the largest man-made floating island. A certificate from the Guinness Book Of World Records was given to TRAM in 1999.

9.3.3   Semi-submersible-type Mega-Float

A Semi-Submersible-type Mega-Float (SSMF) was investigated (Yoshida et al. 2001). SSMF is supported by many column footings or lower hulls and has good response characteristics in waves. A SSMF is expected to be an efficient form of floating structure that can be utilized in open seas, such as the Exclusive Economic Zone of Japan. A typical middle-sized SSMF that could be used as a floating runway and measured 2,200-m long, 300-m wide, and 7-m deep, was used for this research. Figure 9.17 shows a schematic diagram

Figure 9.17 Schematic diagram of semi-submersible type Mega-Float.

of the SSMF. The deck is supported by 320 column footings with a draft of 24 m. The feasibility and optimization of the SSMF have been investigated and validity of the VLFS Oriented Dynamic Analysis Code (VODAC), an analysis code of hydroelastic responses of SSMF, was extensively tested. The computer code was verified by wave tank experiments with a dynamically similar model. Drag forces on multiple columns at critical Reynolds numbers were measured by wind-tunnel tests to study the design of the moorings. Photos of wave tank experiments and wind-tunnel experiments are shown in Figure 9.18. The basic feasibility of the SSMF was confirmed by this research.

9.4 Research and projects after TRAM

9.4.1 Reuse of Phase 2 model

The Phase 2 model was removed from the experimental site in 2001. Some parts of the model were reused for other experiments and commercial uses as shown in Figure 9.19 (homepage of SRCJ). One part of the model was modified as an information technology base. Data-backup experiments, in case of disaster, were performed that utilize the merits of Mega-Float that are minimally influenced by earthquakes. Another application was the World Cup Mega-park. A pre-opening ceremony of world cup soccer in 2002, co-organized by Japan and Korea, was held on the floating structure. More than 13,000 people gathered on the 200-m long and 100-m wide floating

Figure 9.18 Wave tank and wind tunnel experiments.

World Cup Mega-park
L200m x B100m x D2m

Ferry Pier
L143m x B20m x D3m

Marine Park Kumanonada
L120m x B60m x D3m

Figure 9.19 Reuse of Phase 2 model.

Source: Shipbuilding Research Center of Japan.

structure. Other sections of the Mega-Float were used as a ferry pier and marine park for sports fishing.

9.4.2   Extension of Haneda International Airport

Considering the shortage of air transportation capacity in the Tokyo area, the Japanese government decided to construct an additional runway for the

 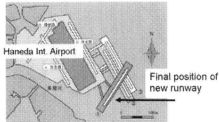

Figure 9.20 Floating runway concept proposed for extension of Haneda International Airport.

Source: Shipbuilders' Association of Japan.

Haneda International Airport (Sato 2003). The runway is planned to be in operation by 2009. Three construction concepts (i.e. floating structure, reclamation and runway-on-piles) were proposed. Based on the results of six years of study by TRAM, the Shipbuilders' Association of Japan proposed the floating type runway as shown in Figure 9.20. The Haneda Airport is located at the mouth of the Tama River and a new runway was to be constructed without disturbing its flow. In order to meet this requirement, the hull design comprises a combination of a pontoon-type VLFS and a rectangular column-pontoon-type semi-submersible hull configuration over its length. The semi-submersible part was adopted for the section extended across the Tama river so that the flow is minimally affected. The service life was determined to be 100 years. In order for the approach course for an airplane not to interfere with the ship lane of Tokyo Bay, the height of the runway was elevated from sea level by 23 m at the northern end and 15 m at the southern end.

Technologies developed for the hull are

- characterization of hydrodynamic and hydroelastic behavior of the new hull;
- extension of design methodology and analysis method developed for pontoon hull to be applied to the new hull configuration; and
- augmentation of wave absorption ability to reduced reflected wave to the navigation area in the port.

Acknowledgments

The author would like to thank the Shipbuilding Research Center of Japan and the Shipbuilders' Association of Japan for permission to use their photos and figures.

References

Armstrong, E. R. (1924) Sea Station, Patent No. 1,511,153, October 7, 1924.

Committee on Realization of Megafloat of Ministry of Land, Infrastructure and Transport (2001) "Safety Evaluation Guideline of Megafloat" (in Japanese).

Fujikubo, M., Yanagihara, D., Matsuda, I., and Olaru, D.V. (2003) "Collapse analysis of a pontoon-type VLFS in waves," 4th Very Large Floating Structures, 185–192.

Hara, S., Ohmatsu, S., Yamakawa, K., Hoshino, K., Yukawa, K., and Toriumi, M. (2003) "Towing field test and critical towing operation on Mega-Float unit," 4th Very Large Floating Structures, 261–270. Homepage of SRCJ, http://www.srcj.or.jp/html/megafloat_en/index.html

Inoue, K., Kinoshita, T., Takagi, K., Terao, H., Okamura, H., Takahashi, M., Esaki, H., and Uehiro, T. (2004) "Development of a Floating Wind Farm with Mooringless System," Proceedings of the Society of Naval Architects of Japan, A-OS7-2 (in Japanese).

Kato, S., Namba, Y., Masanobu, S., and Shimoyama, M. (2001) "Safety Evaluation of Dolphin Mooring System of Hybrid Semisub-Megafloats," 20th International Conference on Offshore Mechanics and Arctic Engineering, OMAE2001/OSU-5016.

Kikutake, K. (1994) "Floating Habitat: A Prospect Marine Cities 1994," 17th Ocean Engineering Symposium, pp. 9–14, 1994 (in Japanese).

Laycock, J. N. (1943) Memorandum to Rear Admiral B. Morell (CEC) USN; Subj: SOCK Operations, 22 December 1943, Ser. No. 01599, Bureau of Yards and Docks, 24 December 1943.

Nelson, S. B. (2001) "Airports Across the Ocean," *Invention & Technology*, Summer, 2001, pp. 32–37, 2001.

Sato, C. (2003) "Results of 6 years research project of Mega-Float," 4th Very Large Floating Structures, 377–383.

Shipbuilders' Association of Japan (2001) "Proposal of Floating Airport for Extension of Haneda Airport" (in Japanese).

Suzuki, H. (2002) "Safety Target of Very Large Floating Structure Used as a Floating Airport," *Marine Structures*, 14: 103–113.

Suzuki, H., Hashimoto, T., and Sekita, K. (2004) "Improvement of Response Characteristics of SPAR-Buoy Type Floating Wind Turbine," *Annual Journal of Civil Engineering in the Ocean*, JSCE, 20: 911–916 (in Japanese).

TRAM (Technological Research Association of Mega-Float) (1998) Redundancy Evaluation of Mooring System.

TRAM (Technical Research Association of Mega-Float) (1999a) Summary of Practical Research on Megafloat Airport in 1999 (in Japanese).

TRAM (Technical Research Association of Mega-Float) (1999b) Technical Guideline of Mega-Float (in Japanese).

TRAM (Technical Research Association of Mega-Float) (1999c) Safety Guideline of Mega-Float (in Japanese).

TRAM (Technical Research Association of Mega-Float) (2001) Summary of Practical Research on Megafloat Airport in 2001 (in Japanese).

TRAM (Technical Research Association of Mega-Float) (2002) Summary of Practical Research on Mega-Float Airport in 2002 (in Japanese).

Yago, K., Ohkawa, Y., Suzuki, H., and Sawai, T. (2003) "A Basic Study on The Floating Wind Power System," 17th Ocean Engineering Symposium, 127–134 (in Japanese).

Yamashita, Y., Yonezawa, M., Shimamune, S., and Kinoshita, Y. (2003) "Joining technology for construction of very large floating structures," 4th Very Large Floating Structures, 229–236.

Yoshida, K., Suzuki, H., Kato, S., Sumiyoshi, H., and Kado, M. (2001) "A Basic Study for Practical Use of Semisub-Megafloat," 20th International Conference on Offshore Mechanics and Arctic Engineering, OMAE2001/OSU5015.

Index

Note: References such as "178–179" indicate (not necessarily continuous) discussion of a topic across a range of pages, whilst "123fig5.1" indicates a reference to figure 5.1 on page 123 and "134tab5.1" a reference to table 5.1 on page 134. Wherever possible in the case of topics with many references, these have either been divided into sub-topics (indented below the main heading) or the most significant discussions of the topic are indicated by page numbers in bold.